T0219636

CAMBRIDGE LIBRARY COLLECTION

Books of enduring scholarly value

Zoology

Until the nineteenth century, the investigation of natural phenomena, plants and animals was considered either the preserve of elite scholars or a pastime for the leisured upper classes. As increasing academic rigour and systematisation was brought to the study of 'natural history', its subdisciplines were adopted into university curricula, and learned societies (such as the London Zoological Society, founded in 1826) were established to support research in these areas. These developments are reflected in the books reissued in this series, which describe the anatomy and characteristics of animals ranging from invertebrates to polar bears, fish to birds, in habitats from Arctic North America to the tropical forests of Malaysia. By the middle of the nineteenth century, this work and developments in research on fossils had resulted in the formulation of the theory of evolution.

Genera of Birds

The naturalist and traveller Thomas Pennant (1726–98) helped popularise British ornithology by meticulously compiling and arranging existing research. At the age of twelve, Pennant had been given Francis Willughby's *Ornithology* (1678), to which he credited his lifelong love of natural history. His own writings on ornithology are heavily based on the classification system devised by Willughby and John Ray, which divides birds primarily into land birds and waterfowl. Although Pennant's brief, accessible book brought few original insights to the field, it boosted public interest in the study and classification of birds. The detailed descriptions of the appearance and habits of each bird are enlivened by the author's elegant turns of phrase. This better-known 1781 version of the 1773 original includes fifteen fine engravings. Pennant's other zoological works include *Arctic Zoology* (1784–5) and his *History of Quadrupeds* (third edition, 1793), both of which are reissued in this series.

Cambridge University Press has long been a pioneer in the reissuing of out-of-print titles from its own backlist, producing digital reprints of books that are still sought after by scholars and students but could not be reprinted economically using traditional technology. The Cambridge Library Collection extends this activity to a wider range of books which are still of importance to researchers and professionals, either for the source material they contain, or as landmarks in the history of their academic discipline.

Drawing from the world-renowned collections in the Cambridge University Library and other partner libraries, and guided by the advice of experts in each subject area, Cambridge University Press is using state-of-the-art scanning machines in its own Printing House to capture the content of each book selected for inclusion. The files are processed to give a consistently clear, crisp image, and the books finished to the high quality standard for which the Press is recognised around the world. The latest print-on-demand technology ensures that the books will remain available indefinitely, and that orders for single or multiple copies can quickly be supplied.

The Cambridge Library Collection brings back to life books of enduring scholarly value (including out-of-copyright works originally issued by other publishers) across a wide range of disciplines in the humanities and social sciences and in science and technology.

Genera of Birds

Thomas Pennant

CAMBRIDGE
UNIVERSITY PRESS

CAMBRIDGE
UNIVERSITY PRESS

University Printing House, Cambridge, CB2 8BS, United Kingdom

Published in the United States of America by Cambridge University Press, New York

Cambridge University Press is part of the University of Cambridge.
It furthers the University's mission by disseminating knowledge in the pursuit of
education, learning and research at the highest international levels of excellence.

www.cambridge.org
Information on this title: www.cambridge.org/9781108067782

© in this compilation Cambridge University Press 2014

This edition first published 1781
This digitally printed version 2014

ISBN 978-1-108-06778-2 Paperback

GENERA

OF

BIRDS.

LONDON.
Printed for B. WHITE
MDCCLXXXI.

ADVERTISEMENT.

THIS trifle was written in the year 1772, and prefented to Doctor ROBERT RAMSAY, Profeffor of Natural Hiftory in *Edinburgh*, for the ufe of the clafs over which he prefided. He printed one impreffion in the following year; and then refigned to me the copy.

DEATH deprived the community of a worthy member, in the lofs of my friend, on *December* 15th, 1778. I fuffer the Dedication to remain in this edition, as a fmall monument to his memory; and of the efteem in which I held a gentleman, ever active in all good offices to

DOWNING,
Dec. 20th, 1780.

THOMAS PENNANT.

ROBERT RAMSAY, M.D.

FELLOW OF THE ROYAL COLLEGE OF PHYSICIANS,

A N D

PROFESSOR OF NATURAL HISTORY IN THE UNIVER-
SITY OF EDINBURGH.

DEAR SIR,

I THINK myfelf happy in having an opportunity of giving you this mark of the fenfe I have of your fteady friendfhip, from its origin, in 1769, to the prefent moment. From the beginning, it has proved a regular feries of good offices : You never confidered me with the jealoufy of a Rival courting the fame Miftrefs; but, with uncommon generofity, promoted all my purfuits after DAME NATURE, whether fhe retired to the depths of the *Highland* Glens, or lurked amidft the intricate groups of the ftormy *Hebrides*. If, in my late expedition, fhe has granted me any favors (for fhe proved rather coy) fhe humbled me by faying, that I was indebted to you for them. So that I find myfelf bound to make public acknowlegements of advantages acquired by means of the clue you gave of arriving at the few I have obtained.

NOT-

DEDICATION.

NOTWITHSTANDING I own your power with the
Lady on your side of the *Tweed*, yet I never can be induced
to omit any opportunity of recommending myself to her
good graces, and, with you, must ever remain a warm ad-
mirer of her universal charms. But the following *analysis*
of one which captivates me most, is now offered to you,
with the hopes of meeting with your approbation, and
that of the several votaries who depend on you for a more
intimate acquaintance with her various beauties. Long
may you enjoy health, and every happiness, to perform so
agreeable a task: May you be successful in extending her
empire: Good fortune attend you in each of her haunts,
whether she affects the air, the woods, or the fields; whether, like an *Oread*, she treads *jocund on the misty moun-
tain's top*; or a *Naiad*, sporting in your rapid streams.
Again, success attend you every where; and may none
but BIRDS of good omen flutter round you.

Sis licet felix ubicunque mavis,
Et memor nostri, mihi care, vivas:
Teque nec lævus vetet ire PICUS,
      Nec vaga CORNIX.

DOWNING,
JAN. 1, 1773.

THOMAS PENNANT.

3

P R E F A C E.

ORNITHOLOGY is a fcience which treats of Birds; defcribes their form, external and internal; and teaches their œconomy and their ufes.

A BIRD is an animal covered with feathers; furnifhed with a bill; having two wings, and only two legs; with the faculty, except in very few inftances, of removing itfelf from place to place through the air.

External Parts of B I R D S.

A BIRD may be divided into HEAD, BODY, and LIMBS.

H E A D,

Roftrum, or bill, is a hard horny fubftance, confifting of an up-   BILL. per and under part, extending from the head, and anfwering to the mandibles in quadrupeds. Its edges generally plain and fharp, like the edge of a knife, *cultrated* *, as the bills of CROWS; but fometimes *ferrated*, as in the TOUCAN; or *jagged*, as in the GAN-

---

This and other terms are explained by figures in the BRITISH ZOOLOGY, vol. i. tab. xv. A few terms are explained from the figure on the title.

NET and some HERONS; or *pectinated*, as in the DUCK; or *denti-culated*, as in the MERGANSERS; but always destitute of real teeth immersed in sockets.

THE base in FALCONS is covered with a naked skin or CERE (CERA;) in some birds with a carneous appendage, as the TUR-KEY; or a callous, as the CURASSO.

IN birds of prey, the bill is hooked at the end, and fit for tear-ing: in CROWS, strait and strong, for picking: in water-fowl, ei-ther long and pointed, for striking; or slender and blunt, for search-ing in the mire; or flat and broad, for gobbling. Its other uses are for building nests; feeding the young; climbing, as in PAR-ROTS; or, lastly, as an instrument of defence, or offence.

NOSTRILS.

(Nares) the nice instruments of discerning their food, are placed either in the middle of the upper mandible, or near the base, or at the base, as in PARROTS; or behind the base, as in TOUCANS and HORNBILLS: but some birds, as the GANNET, are destitute of nos-trils. The nostrils are generally naked, but sometimes covered with bristles reflected over them, as in CROWS; or hid in the fea-thers, as in PARROTS, &c.

PARTS OF THE HEAD.

THE forepart of the head is called the FRONT *(Capistrum;)* the summit *(vertex)* or the crown: the hind part, with the next joint of the neck *(nucha)* the nape: the space between the bill and the eyes, which in HERONS, GREBES, &c. is naked *(lora)* the straps: the space beneath the eyes *(genae)* the cheeks.

ORBITS.

(Orbitae) the eye-lids; in some birds naked, in others covered with short soft feathers.

BIRDS have no eye-brows; but the GROUS kind have in lieu a scarlet naked skin above, which are called *supercilia;* the same

word

word is alfo applied to any line of a different color that paffes from the bill over the eyes.

BIRDS are deftitute of *auricles*, or external ears, having an ori- EARS. fice for admiffion of found, open in all, but OWLS, whofe ears are furnifhed with valves.

THE chin, the fpace between the parts of the lower mandible CHIN. and the neck, is generally covered with feathers; but in the COCK, and fome others, have carneous appendages, called WATTLES *(Palearia;)* in others, is naked, and furnifhed with a POUCH, capable of great dilatation *(Sacculus)* as in the PELICAN and CORVORANTS.

(Collum) the part that connects the head to the body, is longer NECK. in birds than any other animals; and longer in fuch as have long legs than thofe that have fhort, either for gathering up their meat from the ground, or ftriking their prey in the water, except in web-footed fowl, which are, by reverfing their bodies, deftined to fearch for food at the bottom of waters, as SWANS, and the like. Birds, efpecially thofe that have a long neck, have the power of retracting, bending, or ftretching it out, in order to change their center of gravity from their legs to their wings.

B O D Y.

CONSISTS of the BACK *(Dorfum)* which is flat, ftrait, and in- BACK. clines, terminated by the

(Uropygium) furnifhed with two glands, fecreting a fattifh liquor RUMP. from an orifice with which each is furnifhed: and which the birds exprefs with their bills, to oil or anoint the difcompofed parts of their feathers. Thefe glands are particularly large in moft web-footed water-fowl; but in the GREBES, which want tails, they are fmaller.

(Pectus)

BREAST. *(Pectus)* is ridged and very muscular, defended by a forked bone *(clavicula)* the MERRY THOUGHT.

THE short-winged birds, such as GROUS, &c. have their breasts most fleshy or muscular; as they require greater powers in flying than the long-winged birds, such as GULLS, HERONS, which are specifically lighter, and have greater extent of sail.

BELLY. *(Abdomen)* is covered with a strong skin, and contains the entrails.

VENT. THE VENT, or vent-feathers *(Crissum)* which lies between the thighs and the tail. The ANUS lies hid in those feathers.

L I M B S.

WINGS. WINGS *(Alae)* adapted for flight in all birds, except the DODO, OSTRICHES, great AUK, and the PINGUINS, whose wings are too short for the use of flying; but in the DODO and OSTRICH, when extended, serve to accelerate their motion in running; and in the PINGUINS perform the office of fins, in swimming or diving.

BASTARD WING. THE wings have near their end an appendage covered with four or five feathers, called the BASTARD WING *(ala notha)* and *alula spuria.*

LESSER COVERTS. THE lesser coverts *(tectrices)* are the feathers which lie on the bones of the wings.

GREATER COVERTS. THE greater coverts are those which lie beneath the former, and cover the quil-feathers and the secondaries.

QUIL-FEATHERS. THE *Quil-feathers (primores)* spring from the first bones *(digiti* and *metacarpi)* of the wings, and are ten in number.

QUIL-feathers are broader on their inner than exterior sides.

SECONDARIES. THE SECONDARIES *(secondariae)* are those that rise from the second

cond part *(cubitus)* and are about eighteen in number, are equally broad on both fides. The primary and fecondary wing-feathers are called REMIGES.

A TUFT of feathers placed beyond the fecondaries, near the junction of the wings with the body. This, in water-fowl, is generally longer than the fecondaries, and cuneiform. TERTIALS.

THE SCAPULARS are a tuft of long feathers arifing near the junction of the wings *(brachia)* with the body, and lie along the fides of the back, but may be eafily diftinguifhed, and raifed with one's finger. SCAPULARS.

THE INNER COVERTS are thofe that clothe the under fide of the wing. INNER COVERTS.

THE SUBAXILLARY are peculiar to the greater PARADISE. SUBAXILLARY FEATHERS.

THE wings of fome birds are inftruments of offence; the ANHIMA of *Marcgrave* has two ftrong fpines in the front of each wing, a fpecies of Plover, EDW. tab. 47. and 280. has a fingle one on each; the whole tribe of JACANA, and the GAMBO, or fpur-winged Goofe of Mr. *Willughby*, the fame.

THE TAIL is the director, or rudder, of birds in their flight; they rife, fink, or turn by its means; for, when the head points one way, the tail inclines to the other fide: it is, befides, an *equilibrium* or counterpoife to the other parts; the ufe is very evident in the KITE and SWALLOWS. TAIL.

THE TAIL confifts of ftrong feathers *(rectrices)* ten in number, as in the WOODPECKERS, &c. twelve in the HAWK tribe, and many others: the GALLINACEOUS, the MERGANSERS, and DUCK kind, of more.

[T

It is either even at the end, as in moſt birds, or forked, as in Swallows, &c. or cuneated, as in Magpies, &c. or rounded, as in the Purple Jackdaw of *Cateſby*. The Grebe is deſtitute of a tail, the rump being covered with down; and that of the Casso-wary with the feathers of the back.

Immediately over the tail, are certain feathers that ſpring from the lower part of the back, and are called the coverts of the tail (*uropygium.*)

Thighs. *(Femora)* are covered entirely with feathers in all land-birds, except the Bustards and the Ostriches; the lower part of thoſe of all waders, or cloven-footed water-fowl, are naked; that of all webbed-footed fowl the ſame, but in a leſs degree; in rapacious birds, are very muſcular.

Legs. *(Crura)* Thoſe of rapacious fowls very ſtrong, furniſhed with large tendons, and fitted for tearing, and a firm gripe. The legs of ſome of this genus are covered with feathers down to the toes, ſuch as the *Golden Eagle,* others to the very nails; but thoſe of moſt other birds are covered with ſcales, or with a ſkin divided into ſegments, or continuous. In ſome of the Pies, and in all the Passerine tribe, the ſkin is thin and membranous; in thoſe of web-footed water-fowl, ſtrong.

The legs of moſt birds are placed near the center of gravity: In land-birds, or in Waders that want the back toe, exactly ſo; for they want that appendage to keep them erect. Auks, Grebes, Divers, and Pinguins, have their legs placed quite behind, ſo are neceſſitated to ſit erect: Their pace is aukward and difficult, walking like men in fetters; hence *Linnæus* ſtyles their feet *pedes compedes.*

9

THE

THE legs of all cloven-footed water-fowl are long, as they muſt wade in ſearch of food: Of the palmated, ſhort, except thoſe of the FLAMINGO, the AVOSET, and the COURIER.

(Pedes) All land-birds that perch have a large back toe: Moſt of them have three toes forward, and one backward. WOODPECKERS, PARROTS, and other birds that climb much, have two forward, two backward; but PARROTS have the power of bringing one of their hind toes forward while they are feeding themſelves. OWLS have alſo the power of turning one of their fore toes backward. All the toes of the SWIFT turn forwards, which is peculiar among land-birds: The TRIDACTYLOUS WOODPECKER is alſo anomalous, having only two toes forward, one backward: The OSTRICH is another, having but two toes.

FEET.

(Digiti) THE toes of all WADERS are divided; but, between the exterior and middle toe, is generally a ſmall web, reaching as far as the firſt joint.

THE SPOONBILL; and a SANDPIPER I received from *N. America*, have webs that reach half way up each toe, or are *ſemi-palmated*.

THE toes of birds that ſwim are either plain, as in the ſingle inſtance of the common water HEN or GALLINULE; or pinnated, as in the COOTS and GREBES; or entirely webbed or palmated, as in all other ſwimmers.

ALL the PLOVER tribe, or CHARADRII, want the back toe. In the ſwimmers, the ſame want prevales among the ALBATROSSES and AUKS. No water-fowl perch, except certain HERONS; the CORVORANT; and the SHAG.

TOES.

(Ungues) Rapacious birds have very ſtrong, hooked, and ſharp

CLAWS.

sharp claws, VULTURES excepted. Those of all land-birds that roost on trees have also hooked claws, to enable them to perch in safety while asleep.

THE GALLINACEOUS tribe have broad concave claws for scraping up the ground.

GREBES have flat nails like the human.

AMONG water-fowl only the SKUA, *Br. Zool.* II. *No.* 243. and the BLACK TOED GULL, *Br. Zool.* II. *No.* 244. have strong hooked or *aquiline* claws. All land-birds perch on trees, except the STRUTHIOUS and some of the GALLINACEOUS tribe. PARROTS climb; WOODPECKERS creep up the bodies and boughs of trees; SWALLOWS cling.

ALL water-fowl rest on the ground, except certain HERONS, and one species of IBIS, the SPOONBILL, one or two species of DUCKS, and of CORVORANTS.

F E A T H E R S.

FEATHERS are designed for two uses, as coverings from the inclemency of the weather, and instruments of motion through the air. They are placed in such a manner as to fall over one another, *tegulatim*, so as to permit the wet to run off, and to exclude the cold; and those on the body are placed in a quincuncial form, most apparent in the thick-skinned water-fowl, particularly in the DIVERS.

SHAFTS.

THE parts of a feather are, the SHAFTS, corneous, strong, light, rounded, and hollow at the lower part; at the upper, convex above, concave beneath, and chiefly composed of a pith.

On

On each fide the SHAFTS, are the

VANES, broad on one fide, narrow on the other: Each vane   VANES.
confifts of a multitude of thin *laminæ* *, ftiff, and of the nature
of a fplit quil.   Thefe *laminæ* are clofely braced together by
the elegant contrivance of a multitude of fmall briftles; thofe on
one fide hooked, the other ftrait, which lock into each other, and
keep the vanes fmooth, compact, and ftrong.

THE vanes near the bottom of the fhafts are foft, unconnected, and downy.

FEATHERS are of three kinds; fuch as compofe the inftru-   PEN-FEATHERS.
ments of flight; as the PEN-FEATHERS; or thofe which form
the wings and tail, and have a large fhaft.   The vanes of the
exterior fide bending downward, of the interior upwards, lying
clofe on each other,   fo that, when fpread, not a feather miffes
its impulfe on the air †.   The component parts of thefe feathers
are defcribed before.

THE feathers that cover the body, which may be properly   PLUMAGE.
called the PLUMAGE, have little fhaft, and much vane, and never
are exerted or relaxed, unlefs in anger, fright, or illnefs.

THE Down, *Plumæ*, which is difperfed over the whole body   DOWN.
amidft the plumage, is fhort, foft, unconnected, confifts of lanu-
ginous vanes, and is intended for excluding that air or water
which may penetrate or efcape through the former.   This is particularly apparent in aquatic birds, and remarkably fo in the AN-
SERINE tribe.   There are exceptions to the forms of feathers.
The vanes of the fubaxillary feathers of the PARADISE are unconnected, and the *laminæ* diftant, looking like herring-bone.

* *Derham*'s phyfic. theol. 336. tab. f. 18, 19.   † *Derham*.

C   Thofe

Thofe of the tail of the OSTRICH, and head of a fpecies of CU-RASSO, curled. Thofe of the CASSOWARY confift of two fhafts, arifing from 'a common ftem at the bottom. As do, at the approach of winter (after moulting) thofe of the PTARMIGANS of *Arctic* countries. The feathers of the PINGUINS, particularly thofe of the 'wings, confifting chiefly of thin flat fhafts, and more refemble fcales than feathers; thofe of the tail, like fplit whale-bone.

F L I G H T.

THE flight of birds is various; for, had all the fame, none could elude that of rapacious birds. Thofe which are much on wing, or flit from place to place, often owe their prefervation to that caufe: Thofe in the water to diving.

RAPACIOUS. KITES, and many of the FALCON tribe, glide fmoothly through the air, with fcarce any apparent motion of the wings.

PIES. MOST of the order of PIES fly quick, with a frequent repetition of the motion of the wings. The PARADISE floats on the air. WOODPECKERS fly aukwardly, and by jerks, and have a propenfity to fink in their progrefs.

GALLINACEOUS. THE GALLINACEOUS tribe, in general, fly very ftrong and fwiftly; but their courfe is feldom long, by reafon of the weight of their bodies.

COLUMBINE. THE COLUMBINE race is of fingular fwiftnefs; witnefs the flight of the *Meffenger* Pigeon.

PASSERINE. THE PASSERINE fly with a quick repetition of ftrokes; their flight, except in migration, is feldom diftant.

AMONG

Among them, the Swallow tribe is remarkably agile, their evolutions fudden, and their continuance on wing long.

Nature hath denied flight to the Struthious; but ftill, in running, their fhort wings are of ufe, when erect, to collect the wind, and, like fails, to accelerate their motion. Struthious.

Many of the greater Cloven-footed Water-fowl, or Waders, have a flow and flagging flight; but moft of the leffer fly fwiftly, and moft of them with extended legs, to compenfate the fhortnefs of their tails. Rails and Gallinules, fly with their legs hanging down. Waders.

Coots and Grebes, with difficulty are forced from the water; but when they rife, fly fwiftly. Grebes, and alfo Divers, fly with their hind parts downwards, by reafon of the forwardnefs of their wings. Pinnated feet.

Web-footed fowl are various in their flight; feveral have a failing or flagging wing, fuch as Gulls. Pinguins, and a fingle Auk, are denied the power of flight. Wild-geese, in their migrations, do not fly *pell-mell*, but in a regular figure, in order to cut the air with greater eafe; for example, in long lines, in the figure of a ➤ or fome pointed form or letter, as the ancients report that the Cranes affumed, in their annual migrations, till their order was broken by ftorms. Web-footed.

Strymona fic gelidum, bruma pellente, relinquunt,
Poturæ te, *Nile*, Grues, primoque volatu
Effingunt varias, cafu monftrante, figuras,
Mox ubi percuffit tenfas Notus altior alas,
Confufos temerè immiftæ glomerantur in orbes,
Et turbata perit difperfis *litera* * pennis.

Lucan. lib. v. l. 711.

* Υ Δ Λ.

Of

Of the NUPTIALS, NIDIFICATION, and EGGS of BIRDS.

Most birds are monogamous, or pair, in fpring fixing on a mate, and keeping conftant, till the cares of incubation and educating the young brood is paft. This is the cafe, as far as we know, with all the birds of the firft, fecond, fourth, and fifth orders.

BIRDS that lofe their mates early, affociate with others; and Birds that lofe their firft eggs, will pair and lay again. The male as well as female of feveral join alternately in the trouble of incubation, and always in that of nutrition: When the young are hatched, both are bufied in looking out for, and bringing food to the neftlings ; and, at that period, the mates of the melodious tribes, who, before, were perched on fome fprig, and by their warbling alleviated the care of the females confined to the neft, now join in the common duty.

Of the GALLINACEOUS tribe, the greateft part are polygamous, at left in a tame ftate; the PHEASANT, many of the GROUS, the PARTRIDGES, and BUSTARDS, are monogamous; of the GROUS, the COCK of the wood, and the BLACK GAME affemble the females during the feafon of love, by their cries :

Et venerem incertam rapiunt.

THE males of polygamous birds neglect their young, and, in fome cafes, would deftroy them, if they met with them. The œconomy of the STRUTHIOUS order, in this refpect, is obfcure. It is probable that the three fpecies in the genus OSTRICH are polygamous, like the common poultry, for they lay many eggs; the DODO is faid to lay but one.

ALL

ALL Waders, or cloven-footed fowl, are monogamous, except the RUFFS; and all with pinnated feet, as far as I know, are also monogamous.

THE fwimmers, or web-footed fowl, obferve the fame order, as far as can be remarked with any certainty; but many of the AUKS affemble in the rocks in fuch numbers, and each individual fo contiguous, that it is not poffible to determine their method in this article.

IT may be remarked, that the affection of birds to their young, is very violent during the whole time of nutrition, or as long as they continue in a helplefs ftate; but fo foon as the brood can fly and fhift for itfelf, the parents neglect, and even drive it from their haunts, the affection ceafing with the neceffity of it: but, during that period,

The mothers nurfe it, and the fires defend;
The young difmifs'd to wander earth, or air,
There ftops the inftinct, and there ends the care;
The link diffolves, each feeks a frefh embrace,
Another love fucceeds, another race.

N I D I F I C A T I O N.

THE Neft of a bird is one of thofe daily miracles, that, from its familiarity, is paffed over without regard. We ftare with wonder at things that rarely happen, and neglect the daily operations of nature, that ought firft to excite our admiration, and clame our attention.

EACH bird, after nuptials, prepares a place fuited to its fpecies, for the depofiting its eggs, and fheltering its little brood: Diffe-

rent

ient genera, and different fpecies, fet about the tafk in manners fuitable to their feveral natures; yet every individual of the fame fpecies collects the very fame materials, puts them together in the fame form, and chufes the fame fort of fituation for placing this temporary habitation. The young bird of the laft year, which never faw the building of a neft, directed by a heaven-taught fagacity, purfues the fame plan in the ftructure of it, and felects the fame materials as its parent did before. Birds of the fame fpecies, of different and remote countries, do the fame. The SWALLOWS of *England*, and of the remoter parts of *Germany*, obferve the fame order of architecture.

RAPACIOUS. THE nefts of the larger rapacious birds are rude, made of fticks and bents, but often lined with fomething foft. They generally build in high rocks, ruined towers, and in defolate places: enemies to the whole feathered creation, they feem confcious of attacks, and feek folitude. A few build upon the ground.

SHRIKES, the left of RAPACIOUS birds, build their nefts in bufhes, with mofs, wool, &c.

PIES. THE order of PIES is very irregular in the ftructure of their nefts. PARROTS, and, in fact, all birds with two toes forward and two backward (as far as I know) lay their eggs in the hollows of trees. And moft of this order creep along the bodies of trees, and lodge their eggs alfo within them.

CROWS build in trees: Among them, the neft of the MAGPIE, compofed of rude materials, is made with much art, quite covered with thorns, and only a hole left for admittance.

THE nefts of the ORIOLES are contrived with wonderful fagacity, and are hung at the end of fome bough, or between the forks

5 of

of extreme branches. In *Europe*, only three birds have penfile nefts; the common ORIOLE, the PARUS PENDULINUS, or HANG-NEST TITMOUSE, and one more *. But in the Torrid Zones, where the birds fear the fearch of the gliding ferpent and inquifitive monkey, the inftances are very frequent, a marvellous inftinct implanted in them for the prefervation of their young †.

ALL of the GALLINACEOUS and STRUTHIOUS orders lay their eggs on the ground. The OSTRICH is the only exception, among birds, of the want of natural affection: *Which leaveth her eggs in the earth, and warmeth them in the duft, and forgetteth that the foot may crufh them, or the wild beaft may break them.*

GALLINACEOUS.
STRUTHIOUS.

THE COLUMBINE race makes a moft artlefs neft; a few fticks laid acrofs fuffice.

COLUMBINE.

MOST of the PASSERINE order build their nefts in fhrubs or bufhes, and fome in holes of walls, or banks. Several in the Torrid Zone are penfile from the boughs of trees; that of the TAYLOR BIRD ‡, a wondrous inftance. Some of this order, fuch as LARKS, and the GOATSUCKER, on the ground. Some SWALLOWS make a curious plaifter neft beneath the roofs of houfes; and an *Indian* fpecies, one of a certain glutinous matter, which are collected as delicate ingredients for foups of *Chinefe* epicures.

PASSERINE.

MOST of the Cloven-footed Water-fowl, or Waders, lay upon the ground. SPOONBILLS and the COMMON HERON build in trees, and make large nefts with fticks, &c. STORKS build on churches, or the tops of houfes.

WADERS.

COOTS make a great neft near the water-fide.

COOTS.

* Vide Tour in *Scotland*, 2d Ed. page 101.
† Indian Zool.
‡ The fame.

GREBES

GREBES.

GREBES in the water, a floating neft, perhaps adhering to fome neighboring reeds.

WEB-FOOTED.

WEB-FOOTED fowl breed either on the ground, as the AVOSET, TERNS, fome of the GULLS, MERGANSERS, and DUCKS: the laft pull the down from their breafts, to make a fofter and warmer bed for their young. AUKS and GUILLEMOTS lay their eggs on the naked fhelves of high rocks; PINGUINS in holes under ground: Among the PELICANS, that which gives name to the genus makes its neft in the defart, on the ground. SHAGS, fometimes on trees; CORVORANTS and GANNETS, on high rocks, with fticks, dried ALGÆ, and other coarfe materials.

E G G S.

RAPACIOUS.

RAPACIOUS birds, in general, lay few eggs; EAGLES, and the larger kinds, fewer than the leffer. The eggs of FALCONS and OWLS are rounder than thofe of moft other birds.

PIES.

THE order of PIES vary greatly in the number of their eggs.

PARROTS lay only two or three white eggs.

CROWS lay fix eggs, greenifh, mottled with dufky.

CUCKOOS, as far as I can learn, two.

WOODPECKERS, WRYNECK, and KINGSFISHER, lay eggs of a moft clear white and femi-tranfparent color. The WOODPECKERS lay fix, the others more.

THE NUTHATCH lays often in the year, eight at a time, white, fpotted with brown.

THE HOOPOE lays but two cinerous eggs.

THE CREEPER lays a great number of eggs.

THE HONEYSUCKER, the leaft and moft defencelefs of birds, lays

but

but two: but Providence wisely prevents the extinction of the genus, by a swiftness of flight that eludes every pursuit.

THE GALLINACEOUS order, the most useful of any to mankind, lay the most eggs, from eight to twenty; *Benigna circa hoc natura, innocua et esculenta animalia fœcunda generavit*, is a fine observation of Pliny. With exception to the BUSTARD, a bird that hangs between the GALLINACEOUS and the WADERS, which lays only two.

GALLINACEOUS.

THE COLUMBINE order lays but two white eggs; but the domestic kind, breeding almost every month, supports the remark of the *Roman* naturalist.

ALL of the PASSERINE order lay from four to six eggs, except the Titmice and the Wren, which lay fifteen or eighteen, and the Goatsucker, which lays only two.

PASSERINE.

THE STRUTHIOUS order, which consists but of two genera, disagree much in the number of eggs: the OSTRICH laying many, as far as fifty; the DODO but one.

STRUTHIOUS.

THE CLOVEN-FOOTED Water-fowl, or WADERS, lay, in general, four eggs. The CRANE and the NORFOLK PLOVER seldom more than two. All those of the SNIPE and PLOVER genus are of a dirty white, or olive, spotted with black, and scarcely to be distinguished in the holes they lay in. The bird called the LAND RAIL (an ambiguous species) lays from fifteen to twenty. Of birds with pinnated feet, the COOT lays seven or eight eggs, and sometimes more. GREBES from four to eight, and those white.

WADERS.

THE WEB-FOOTED, or Swimmers, differ in the number of their eggs. Those which border on the order of Waders, lay few eggs; the AVOSET, two; the FLAMINGO, three; the ALBATROSS, the AUKS, and GUILLEMOTS, lay only one egg apiece: the eggs of

WEB-FOOTED.

D the

the two laſt, are of a ſize ſtrangely large in proportion to the bulk of the birds. They are commonly of a pale green color, ſpotted and ſtriped ſo variouſly, that not two are alike; which gives every individual the means of diſtinguiſhing its own, on the naked rock, where ſuch multitudes aſſemble.

DIVERS, only two.

TERNS and GULLS lay about four eggs, of a dirty olive, ſpotted with black.

DUCKS lay from eight to twenty eggs; the eggs of all the genus are of a pale green, or white, and unſpotted.

PINGUINS lay two eggs *; white, and remarkably round.

OF the PELICAN *genus*, the GANNET lays but one egg; the SHAGS, or CORVORANTS, ſix or ſeven, all white; the laſt the moſt oblong of eggs

A MINUTE account of the Eggs of birds, merits a treatiſe of it-ſelf, or ſhould follow the deſcription of each ſpecies. This is only meant to ſhew the great conformity nature obſerves in the ſhape and colors of the eggs of congenerous birds; and alſo, that ſhe keeps the ſame uniformity of color in the eggs, as in the plumage of the birds they belong to.

Zinanni publiſhed, at *Venice*, in 1737, a treatiſe on eggs, illuſ-trated with accurate figures of 106 eggs. Mr. *Reyger* of *Dant-zick* publiſhed, in 1766, a poſthumous work by *Klein*, with 21 plates, elegantly coloured: But much remains for future writers.

S Y S T E M.

CONSIDERING the many fyftems that have been offered to the public of late years *, I hope I fhall not be accufed of national partiality, in giving the preference to that compofed by Mr. RAY in 1667, and afterwards publifhed in 1678. It would be unfair to conceal the writer, from whom our great countryman took the original hint of forming that fyftem, which has fince proved the foundation of all that has been compofed fince that period.

IT was a *Frenchman*, BELON of *Mans*, who firft attempted to range birds according to their natures, and performed great matters, confidering the unenlightened age he lived in; for his book was publifhed in 1555. His arrangement of rapacious birds is as judicious as that of the lateft writers, for his fecond chapter treats of VULTURES, FALCONS, SHRIKES, and OWLS; in the two next, he paffes over to the Web-footed Water-fowl, and to the Cloven-footed; in the fifth, he includes the GALLINACEOUS and STRUTHIOUS, but mixes with them the PLOVERS, BUNTINGS, and LARKS; in the fixth are the PIES, PIGEONS, and THRUSHES; and the feventh takes in the reft of the PASSERINE order.

NOTWITHSTANDING the great defeats that every naturalift will at once fee in the arrangement of the leffer birds of this writer, yet he will obferve a rectitude of intention in

* By *M. Barrere* of *Perpignan* in 1745, *Mr. Klein* in 1750, *Mr. Moehring* in 1753, *M. Briffon* in 1760, and by *Linnæus* at different periods. Mr. RAY formed (in conjunction with Mr. WILLUGHBY) his tables of animals, in the winter 1667, for the ule of Bifhop WILKIN's *real character.*

general, and a fine notion of fyftem, which was left to the
following age to mature and bring to perfection. Accord-
ingly, Mr. RAY, and his illuftrious pupil the Hon. FRA.
WILLUGHBY, affumed the plan; but, with great judgment,
flung into their proper ftations and proper genera, thofe which
BELON had confufedly mixed together. They formed the great
divifion of TERRESTRIAL and AQUATIC birds; they made every
fpecies occupy their proper place, confulting at once exterior
form, and natural habit. They could not bear the affected
intervention of aquatic birds in the midft of terreftrial birds:
They placed the laft by themfelves, clear and diftinct from thofe
whofe haunts and œconomy were fo different.

VARIOUS attempts have been made to alter this fyftem of
our countrymen. It is a difagreeable and invidious tafk to
expofe the defects of other methodifts, who may have, in many
refpects, great merit. I leave that to the peevifh malignancy of
the minute critics; therefore fhall only acknowlege the fources
from which I draw the materials of the prefent work, and give
each their due fhare of merit.

Mr RAY's general plan is fo judicious, that to me it feems
fcarcely poffible to make any change in it for the better; yet,
notwithftanding he was in a manner the founder of fyftematic
Zoology, later difcoveries have made a few improvements on his
labors. My candid friend LINNÆUS did not take it amifs, that
I, in part, neglected his example; for I permit the LAND-
FOWL to follow one another, undivided by the WATER-FOWL,
the *Grallæ* and *Anferes* of his fyftem; but, in my generical arrangement, I moft punctually attend to the order he has given
in

in his feveral divifions, except in thofe of his *Anferes,* and a few
of his *Grallæ.* For, after the manner of *M. Briffon,* I make a
diftinct order of WATER-FOWL with pinnated feet, placing them
between the WADERS or CLOVEN-FOOTED Water-fowl and the
Web-footed. The OSTRICH, and Land-birds with wings ufelefs
for flight, I place as a diftinct order. The TRUMPETER *(Pfophia Linnæi)* and the BUSTARDS, I place at the end of the GAL-
LINACEOUS tribe. All are Land-birds. The firft *multiparous,*
like the generality of the GALLINACEOUS tribe; the laft granivorous, fwift runners, avoiders of wet-places; and both have
bills fomewhat arched. It muft be confeffed, that both have
legs naked above the knees; and the laft, like the WADERS, lay
but few eggs. They feem ambiguous birds that have affinity
with each order; and it is hoped, that each naturalift may be
indulged the toleration of placing them as fuits his own opinion.
Before I conclude, let me not pafs over the affiftance received in
fome of my definitions from Mr. SCOPOLI, an ornithologift of *Car-
niola,* who, in 1768, favored the world with a moft elaborate account of the birds that had fallen within his obfervation. Thus,
I flatter myfelf, I have given every naturalift, I am indebted
to, his due.

—— *Miferum eft aliorum incumbere famæ.*
Ne collapfa ruant fubductis tecta Columnis.

T A B L E

TABLE of ARRANGEMENT, with the correspondent ORDERS and GENERA in the SYSTEMA NATURÆ of LINNÆUS.

DIVISION I. LAND-BIRDS.   DIV. II. WATER-FOWL.

	Order I. Rapacious.	Accipitres LINNÆI.
	II. Pies.	Picæ.
	III. Gallinaceous.	Gallinæ.
Division I.	IV. Columbine.	Passeres.
	V. Passerine.	Passeres.
	VI. Struthious.	{ Gallinæ. Grallæ.

	OrderVII. Cloven-footed, or Waders.	} Grallæ.
Division II.	VIII. Pinnated feet.	{ Anseres. Grallæ.
	IX. Web-footed.	{ Anseres. Grallæ.

DIV. I.

ORDER I. RAPACIOUS.

1 Vulture	Vultur	3 Owl	Strix
2 Falcon	Falco		

ORDER II.

ORDER II. PIES.

4 Shrike	Lanius	17 Curucui	Trogon
5 Parrot	Pſittacus	18 Barbet	Bucco
6 Toucan	Ramphaſtos	19 Cuckoo	Cuculus
7 Motmot	Ramphaſtos	20 Wryneck	Junx
8 Hornbill	Buceros	21 Woodpecker	Picus
9 Beefeater	Buphaga	22 Jacamar	Alcedo
10 Ani	Crotophaga	23 Kingsfiſher	Alcedo
11 Wattle		24 Nuthatch	Sitta
12 Crow	Corvus	25 Tody	Todus
13 Roller	Coracias	26 Bee-eater	Merops
14 Oriole	Oriolus	27 Hoopoe	Upupa
15 Grakle	Gracula	28 Creeper	Certhia
16 Paradiſe	Paradiſæa	29 Honeyſucker	Trochilus

ORD. III. GALLINACEOUS.

30 Cock	Phaſianus	35 Pheaſant	Phaſianus
31 Turkey	Meleagris	36 Grous	Tetrao
32 Pintado	Numida	37 Partridge	Tetrao
33 Curaſſo	Crax	38 Trumpeter	Pſophia
34 Peacock	Pavo	39 Buſtard	Otis

ORD. IV. COLUMBINE.

40 Pigeon   Columba

ORD. V.

Ord. V. PASSERINE.

41 Stare	Sturnus	49 Flycatcher	Muſcicapa
42 Thruſh	Turdus	50 Lark	Alauda
43 Chatterer	Ampelis	51 Wagtail	Motacilla
44 Coly	Loxia	52 Warblers	Motacilla
45 Groſbeak	Loxia	53 Manakin	Pipra
46 Bunting	Emberiza	54 Titmouſe	Parus
47 Tanager	Tanagra	55 Swallow	Hirundo
48 Finch	Fringilla	56 Goatſucker	Caprimulgus

Ord. VI. STRUTHIOUS.

57 Dodo	Didus	58 Oſtrich	Struthio

DIV. II.

Ord. VII. CLOVEN-FOOTED, or WADERS.

59 Spoonbill	Platalea	68 Sandpiper	Tringa
60 Screamer	Palamedea	69 Plover	Charadrius
61 Jabiru	Myƈteria	70 Oyſtercatcher	Hæmatopus
62 Boatbill	Cancroma	71 Jacana	Parra
63 Heron	Ardea	72 Pratincole	Hirundo
64 Umbre	*Scopus Briſſ.*	73 Rail	Rallus
65 Ibis	Tantalus	74 Sheath-Bill	
66 Curlew	Scolopax	75 Gallinule	Fulica
67 Snipe	Scolopax		

Ord. VIII.

ORD. VIII. PINNATED-FEET.

76 Phalarope	Tringa	78 Grebe	Colymbus
77 Coot	Fulica		

ORD. IX. WEB-FOOTED.

79 Avoſet	Recurviroſtra	88 Gull	Larus
80 Courier	*Currira Briſſ.*	89 Petrel	Procellaria
81 Flammant	Phœnicopterus	90 Merganſer	Mergus
82 Albatroſs	Diomedea	91 Duck	Anas
83 Auk	Alca	92 Pinguin	⎰ Diomedea ⎱ Phæton
84 Guillemot	Colymbus		
85 Diver	Colymbus	93 Pelican	Pelicanus
86 Skimmer	Rhyncops	94 Tropic	Phæton
87 Tern	Sterna	95 Darter	Plotus

E   Explanation

Explanation of the Figure on the Title-Page.

1 Baftard wing, *Alula fpuria.*
2 Leffer coverts of the wings, *Tectrices primæ.*
3 Greater coverts, *Tectrices fecundæ.*
4 Quill feathers, *Primores.*
5. Secondary feathers, *Secundariæ.*
6 Tertials.
7 Coverts of the tail, *Uropygium.*
8 Vent feathers, *Criffum.*
9 Tail feathers, *Rectrices.*

DIV. I. LAND-FOWL.

ORDER I.

RAPACIOUS.

ACCIPITRES *Linnæi.*

BILL, ftrait, hooked only at the end; edges cultrated, bafe co-   I. VULTURE.
    vered with a thin fkin.

NOSTRILS, differing in different fpecies.

TONGUE, large and flefhy.

HEAD, cheeks, chin, and often neck, either naked or covered
    only with down or fhort hairs; the neck retractile.

CLAW, often hanging over the breaft.

LEGS and FEET, covered with great fcales; the firft joint of the
    middle toe connected to that of the outmoft, by a ftrong
    membrane.

CLAWS, large, little hooked, and very blunt.

INSIDES of the wings covered with down.

King of the Vultures. Bearded and crefted Vultures. EDW. II.   EXAMPLE.
    CVI. CCXI. Bengal and Secretary Vultures. Latham's *Syn.*
    of Birds. Pl. 1, 2.

The γυψ of *Ariftotle,* who mentions two fpecies.

Vultur of *Linnæus,* genus I. who enumerates VIII fpecies. The
    Vultur and *Vautour* of *Briffon,* who defcribes XII fpecies. *M.*
    de Buffon VIII. Mr. *Ray* VIII.

RAPACIOUS.

No Vultures north of the *Baltic*, none in *Great Britain*. Various
ſpecies in *Europe*, *Aſia*, *Africa*, and *America*, as low as *Terra
del Fuego*.

A ſluggiſh, ungenerous race; prey oftener on dead animals, and
even putrid carcaſes, than on living creatures. Their ſenſe of
ſmelling moſt exquiſite. Collect in flocks from afar; directed
to their prey by the ſagacity of their noſtrils. Fly ſlowly and
heavily. Are moſt greedy, and voracious to a proverb. Are
not timid, for they prey in the midſt of cities, undaunted by
mankind.

II. FALCON.   BILL, hooked; covered at the baſe with a naked membrane, or
cere.

NOSTRILS, ſmall, oval, placed in the cere.

TONGUE, large, fleſhy, and often cleft at the end.

HEAD and NECK, covered with feathers.

LEGS and FEET, ſcaly; middle toe connected, from its firſt joint,
to that of the outmoſt, by a ſtrong membrane.

CLAWS, large, much hooked, and very ſharp; that of the outmoſt toe the leſt.

The FEMALE larger and ſtronger than the male.

EXAMPLE.   Golden Eagle, *Br. Zool. fol. tab.* A. Falcon gentil. *Br. Zool. I.
tab.* XXI. Chineſe. E. N. Zealand, F. Latham's *Syn. of Birds.*
Pl. 3, 4.

A carnivorous, rapacious race; not gregarious; quick-ſighted:
Generally fly high. Build in lofty places; except a few ſpecies which neſtle on the ground.

Eagles and the larger kind of Falcons do not lay more than four
eggs; ſome of the leſſer, ſuch as the *Keſtril*, lay ſix or ſeven;

the

CRESTED HOBBY.

the Eagles, properly fo called, feldom more than two or three:
Drink feldom; the juices of their animal-food preventing
thirft. Capable of enduring very long abftinence. Very long
lived. Are clamorous; their note puling and plaintive.
Strike their prey with their feet. Their excrements white and
fluid. Vomit up the indigefted hair or feathers of their prey,
in form of a round ball. Vary in the color of their plumage
at different ages; fo the fpecies are often unneceffarily multi-
plied by Ornithologifts. Inhabit every climate.

Mr. *Ray* and *M. Briffon* feparate the Eagles from the Falcons.
The firft has VIII fpecies of Eagles, and XXV of Falcons or
Hawks. The laft, XV of Eagles, and XXXVII of Falcons.
Linnæus, who, with much propriety, places both in one genus,
enumerates thirty-two. Mr. *Ray*'s divifion of the fluggifh,
and of the more active and generous, a very judicious one.

Bill, hooked; bafe covered with briftles; no cere.   III. Owl.
Nostrils, oblong.
Tongue, cleft at the end.
Eyes, very large and protuberant, furrounded by a circle of
    feathers.
Head, very large and round; full of feathers.
Ears, large and open.
Outmost Toe, verfatile, or capable of being turned back, fo
    as to act with the back toe.
Claws, hooked and fharp.

Eagle Owl, *Br. Zool.* I. *tab.* XXIX. Owls. Latham's *Syn. of*   Ex.
    Birds. Pl. 5.
A nocturnal Bird, preys in the evening and by night; often flies
                                along

along the ground in fearch of prey; carnivorous; quick of hearing; winks in the day; makes a hooting noife in the night; fometimes a fqueaking. Snores loud. Builds in rocks, in hollow trees, or ruined edifices. Lays not more than five eggs. Inhabits every climate.

Mr. *Ray* divides this genus into two; thofe with and thofe without Horns; enumerates III fpecies with, and VIII without. *M. Briffon* ftyles the firft *Afio*, and has IX; the other *Strix*, and has XI fpecies. *Buffon* XV.

ORDER

ORDER II.

P I E S.

P I C Æ *Linnæi.*

BILL, ftrait, hooked only at the end ; near the end of the upper   IV. SHIRKE.
mandible a fharp procefs. No cere.

NOSTRILS, round, covered with ftiff briftles.

TONGUE, jagged at the end.

TOES divided to the origin.

TAIL cuneiform.

Butcher Birds, *Br. Zool. fol. tab.* C. I. *Br. Zool.* I. *tab.* XXXIII.   Bx.
Carnivorous or infectivorous ; kill fmall birds by ftrangling,
or by crufhing their fkull with their bills, then pull them to
pieces, and ftick the fragments on thorns ; do the fame by infects. Bold, noify, and querulous. Build in low bufhes.
Lay fix eggs.

The genus that connects the rapacious Birds and Pies ; agreeing with the firft in the ftrength and crookednefs of the bill,
and its predatory life ; with the laft, in the form of the toes,
the tongue, and tail. Nearly related to the *Magpie :* The
French ftyle it *Pie-Griefche.*

Different fpecies found in the new and old world, and in all climates, except within the *Arctic* circle.

The Butcher Birds or Skrikes of Mr. *Ray,* who defcribes IV
fpecies. The *Lanius* of LINNÆUS, who has XXVI fpecies.

                                    The

The *Lanius* and *Pie-Griefche* of *Briffon*, who reckons up XXVI. *Buffon* XIV.

I reject the compound name of *Butcher-Bird*, and retain the old *Englifh* name of *Shrike*, from the noife.

V. PARROT.   BILL, hooked from the bafe : Upper mandible moveable.

NOSTRILS, round, placed in the bafe of the bill.

TONGUE, broad, blunt at the end.

HEAD, large ; crown flat.

LEGS, fhort. TOES, two backward, two forward.

Ex.   Maccaw EDW. CLVIII. Parrot, CLXVI.

Gregarious, clamorous ; the wild note loud and harfh. Very docile, imitative of founds ; imitates the human fpeech. Climbs by help of the bill and feet. Makes ufe of the feet as hands to convey meat to the mouth, turning the legs outward. Frugivorous : Can crack the hardeft kernels. Breeds in hollow trees. Makes no neft : Lays two or three white eggs : Inhabits within the Tropics, *Africa, Afia,* and *America* ; a few are found as far North as *Carolina :* and South as the Straits of *Magellan.*

Pfittacus of LINNÆUS, and *Briffon,* IV. 182. The firft has XLVII. the laft XC·fpecies.

VI. TOUCAN.   BILL, moft difproportionably large ; convex and carinated at top, and bending at the end ; hollow ; very light, ferrated at the edges.

NOSTRILS, fmall and round, placed clofe to the head, and hid in the feathers.

TONGUE, long and narrow, feathered at the edges.

TOES, two forward, and two backward.

Toucans

WHITE COLLARED PARROT.

Toucans Edw. LXIV. CCXXXIX.   Ex.

A genus confined to *America*, within the Tropics. Feeds on
fruits: Breeds in hollow trees. Is very noify; eafily made
tame.

Mr *Ray*, mifled by the name of *Brafilian Pie*, places it with the
Magpie. Linnæus calls it *Rhamphaftos*, a Ῥαμφος, a broad
fword, from the form of its bill, and has VIII fpecies. *Briffon*,
IV. 407. retains the *Brafilian* name *Toucan*, and has XII
fpecies.

Bill, ftrong, flightly incurvated; ferrated at the edges.   VII. MOTMOT.
Nostrils, covered with feathers.

Tail, cuneiform: The two middle feathers much longer than
the others: Near the ends quite deftitute of webs. The webs
at the ends fubovated.

Toes, three before, one behind; the fore toes clofely united al-
moft their whole length.

Brafilian faw-billed Roller. *Edw.* CCCXXVI.   Ex.
Inhabits *S. America.*

Ramphaftos Momota of Linnæus.

Momotus of *Briffon*, IV. 464. who has II fpecies. I retain the
Mexican name in *Fernandez hift. av. Nov. Hifp.* 52.

Great bending Bill, oft-times a large protuberance refembling   VIII. HORN-
another bill on the upper mandible. Edges jagged.   BILL.
Nostrils, fmall, round, placed behind the bafe of the bill.
Tongue.

Legs, fcaly : Toes, three forward, one backward : The middle
    connected to the outmoft, as far as the third joint; to the in-
    moft, as far as the firft.

Ex.   Several Bills Edw. CCLXXXI. *Wil. orn. tab.* XVII. A fpecies
    with a horn pointing forward, and wattles under the chin, en-
    graven in *Moore*'s travels into the inland parts of *Africa*, p. 108.
    Found in the *Indian* iflands.
    Buceros of Linnæus, a βους an ox, and κερας a horn, from the form
    of the bill. *Hydrocorax* of *Briffon*, II. 565, or Water Raven,
    from its being fuppofed to inhabit watry places.
    Linnæus has IV fpecies. *Briffon* V.

IX. BEEF-   Bill, ftrong, thick, ftrait, nearly fquare. Upper mandible a little
EATER.   protuberant; on the lower, a large angle.
    Tongue.
    Toes, three before, one behind. The middle connected to the
    outmoft as far as the firft joint.

Ex.   Le pique Bœuf. *Briffon* II. *tab.* XLII.
    Inhabits *Senegal*. Only one known fpecies.
    Buphaga of Linnæus and *Briffon* II. 437. a βους an ox, and φαγειν
    to eat, becaufe it picks holes in the backs of cattle, to get at
    the *Larvæ* of infects depofited there.

X. ANI.   Bill, compreffed, greatly arched, half oval, thin, cultrated at top.
    Nostrils, round.
    Toes, two backward, two forward.
    Ten feathers in the Tail.

                                                              Razor-

Razor-bill'd Blackbird. *Catefby Carol. app.* III. the feet faultily   Ex.
   expreffed. Le Bout de Petun, *Briffon* IV. *tab.* XVIII.
Inhabits *South America :* Within the Tropics.
Crotophagus of LINNÆUS and *Briffon* IV. 177. from Κρoῖων, becaufe
   this genus feeds on ticks. Only II fpecies. Mr. *Ray* places it
   at the end of the Parrots. I retain the *Brafilian* name *Ani.*

BILL, ftrong, thick, rounded at top; convex.   XI. WATTLE.
NOSTRILS, covered partly above with a flap; and near their ends
   with a tuft of feathers : On each fide of the bafe of the bill, a
   red, thin, flefhy membrane, or Wattle, of a round form.
TONGUE, truncated, fplit, culiated.
TAIL, long and cuneiform.
LEGS and FEET, ftrong: the firft carinated behind.
TOES, large; three forward, two backward. CLAWS, great and
   crooked, efpecially that of the hind toe.
A non-defcript genus, as yet difcovered only in *New Zeland.*

BILL, ftrong, upper mandible a little convex. Edges cul-   XII. CROW.
   trated.
NOSTRILS, covered with briftles reflected over them.
TONGUE, divided at the end.
TOES, three forward, one backward, the middle joined to the out-
   moft as far as the firft joint.

Royfton Crow, *Br. Zool. fol. tab.* D. I.   Ex.
Different fpecies found in every climate : clamorous : promifcuous
   feeders : build in trees : lay about fix eggs.
Corvus of LINNÆUS, who mentions XIX fpecies.
Briffon divides this genus into *Coracias,* or the Chough ; *Corvus,*

or Crow; *Pica*, or Magpye; *Garrulus*, or Jay; *Nucifraga*, or Nutbreaker; including XXIII fpecies.

XIII. ROLLER.   BILL, ftrait, bending a little towards the end, edges cultrated.

NOSTRILS, narrow and naked.

TOES, three forward; divided to thei origin; one backward.

Ex.   Blue Jay *Edw.* CCCXXVI. Roller *Br. Zool.* II. *App. Europe, Afia, Africa,* and the hot parts of *Ameriſa.*   A genus nearly related to the *Crow.*   Thence LINNÆUS calls it *Coracias:* a word of *Ariftotle*'s, applied only to what we call the *Cornifh Chough.* Κοραχιας Φοινιχορυγχος, *Hift. an. lib.* IX. *c.* 24.

Coracias of LINNÆUS, who has VI fpecies.   *Galgulus* of *Briffon,* who has X fpecies.

XIV. ORIOLE.   BILL, ftrait, conic, very fharp pointed, edges cultrated, inclining inwards.   Mandibles of equal length.

NOSTRILS, fmall, placed at the bafe of the bill, and partly covered.

TONGUE, divided at the end.

TOES, three forward, one backward: the middle joined near the bafe to the outmoft one behind.

Ex.   Redwing Starling *Catefby Carol.* 1. XIII.

In general, inhabitants of *America.*

A numerous race, gregarious, noify, frugivorous, granivorous, voracious: often have penfile nefts.

LINNÆUS enumerates, under the title of *Oriolus,* XX fpecies, but fome belong to the *Turdine* or Thrufh kind.   *Briffon* II. 85. calls this genus *Ifterus,* and has XXX fpecies.   The genuine *Oriolus* is a Thrufh.

BILL,

BILL, convex, thick, compreſſed a little on the ſides, cultrated. **XV. GRAKLE.**

NOSTRILS, ſmall, near the baſe of the bill; often near the edge.

TONGUE, entire; rather ſharp at the end.

TOES, three forward, one backward; the middle connected at the baſe to the outmoſt.

CLAWS, hooked and ſharp.

Mino, EDW. XVII. Chineſe Starling. EDW. XIX.   Ex.

Inhabits *Aſia* and *America*.

Gracula of LINNÆUS, VIII ſpecies. *Icterus*, *Pica*, and *Turdus* of *Briſſon*.

None of LINNÆUS's ſpecies can be the *Graculus* of *Pliny*, or our Chough. For all his are *Aſiatic*, *African*, or *American*.

BILL, ſlightly bending. The baſe covered with velvet-like feathers. **XVI. PARADISE.**

NOSTRILS, ſmall, and concealed by the feathers.

TAIL, conſiſting of ten feathers; two very long naked ſhafts, ſpringing from above the rump.

LEGS and FEET, very large and ſtrong; three toes forward, one backward: the middle connected as far as the firſt joint of the exterior.

CLAWS, large, hooked, and ſharp.

Birds of Paradiſe, EDW. CX. CXI.   Ex.

Floats on the air, and often flies ſwiftly backwards and forwards, like the Swallow; often lights, and perches on trees; feeding on fruits, and even ſmall birds.

Inhabits *New Guinea* and the *Molucca Iſles*. *Paradiſæa* of LINNÆUS, III ſpecies. *Manucodiata* of *Briſſon* II. 130. only II ſpecies. More ſince diſcovered.

BILL,

XVII. CURUCUI BILL, fhort, thick, and convex.

NOSTRILS, covered with ftiff briftles.

TONGUE.

TOES, two backward, two forward.

LEGS, feathered down to the toes.

TAIL, confifts of twelve feathers

Ex.   Yellow-bellied green Cuckow, EDW. CCCXXXI.

Fafciated Couroucou.   *Ind. Zool. tab.* V.

Probably have the manners of the Woodpeckers.

Inhabits *South America.*

Trogon of LINNÆUS, III fpecies.   The fame of *Briffon,* IV. 164.
has VI fpecies.   The reafon for the name *Trogon* feems to be,
becaufe *Pliny* has fuch a name after the *Picus.*   As the genus
is *Brafilian,* I retain the name of the country.

XVIII. BARBET.   BILL, ftrong, ftrait, bending a little towards the point.   Bafe covered with ftrong briftles, pointing downwards.

NOSTRILS, hid in the feathers.

TONGUE.

TOES, two backward, two forward, divided to their origin.

TAIL, confifting of ten weak feathers.

Ex.   Yellow Woodpecker, with red fpots, EDWARDS, CCCXXXIII.

Inhabits *South America* and the *Indian Iflands.*

Bucco of LINNÆUS and *Briffon* IV. 91.   The firft has I. fpecies.
The laft V.   *Briffon* ftyles it *Bucco* from the fulnefs of the
cheeks; *Barbu* from its briftles, a fort of beard, from which I
form the generical name *Barbet.*

5

BILL, weak, a little bending.   XIX. CUCKOO.
NOSTRILS, bounded by a fmall rim.
TONGUE, fhort, pointed.
TOES, two forward, two backward.
TAIL, cuneated; confifts of ten foft feathers.

Cuckoo, *Br. Zool. fol. tab.* G. G. I. *Br. Zool.* I. *tab.* XXXVI.   Ex.
Inhabits every climate.
Cuculus of LINNÆUS and *Briffon*, IV. 104. The one has XXII
    fpecies; the other XXVIII.
The κοκκυξ, and *Coccyx* of the *Ancients*, a word formed from the
    found of the *European* fpecies. *Cuculus* is only ufed in an op-
    probrious fenfe.

BILL, weak, flender, pointed.   XX. WRYNECK.
NOSTRILS, large and oval, near the ridge of the bill.
TONGUE, very long, cylindric, very flender, and terminated by a
    hard point, miffile.
TOES, two forward, two backward.
TAIL, confifting of ten even and foft feathers.

Wryneck, *Br. Zool. fol. tab.* G. *Br. Zool.* I. *tab.* XXXVI.   Ex.
Its manners, *vide Br. Zool.*
Inhabits *Europe* and *Bengal.* Only one fpecies known. Ιυγξ of
    Ariftotle, *Jynx* of *Pliny*, LINNÆUS, and *Briffon*, vol. iv. 3.

BILL, ftrait, ftrong, angular; cuneated at the end.   XXI. WOOD-
NOSTRILS, covered with briftles reflected down.   PECKER.
TONGUE, very long, flender, cylindric, bony, hard, and jagged at
    the end, miffile.

                                 TOES,

Toes, two forward, two backward.

Tail, confifting of ten hard, ftiff, fharp-pointed feathers.

Ex.

Woodpeckers, *Br. Zool. fol. tab.* E. *Br. Zool.* I. *tab.* XXXVII.

The manners, *vide Br. Zool.*

Inhabits all the Continents.

Δρυοχολαπ𝑙ης or Oak-rapper of *Ariftotle, Picus Martius* of *Pliny, Picus* of Linnæus and *Briffon,* IV. 8. Linnæus has XXI. *Briffon* XXXI fpecies.

XXII. JACAMAR

Bill, long, ftrait, fharp pointed, quadrangular.

Nostrils.

Tongue, fhort.

Legs, feathered before to the Toes.

Toes, difpofed two forward, two backward. The two foremoft clofely connected together.

Ex.

Jacamiciri Edw. CCCXXXIV.

Inhabits *S. America.*

Alcedo Galbula of Linnæus. *Galbula* of *Briffon,* IV. 86. who has II fpecies. I retain his name from the *Brafilian Jacamiciri.*

XXIII. KINGS-
FISHER.

Bill, long, ftrong, ftrait, fharp pointed.

Nostrils, fmall, and hid in the feathers.

Tongue, fhort, broad, fharp pointed.

Legs, fhort; three toes forward, one backward: three lower joints of the middle toe joined clofely to thofe of the outmoft.

Ex.

Kingsfifher, *Br. Zool. fol. tab.* I. *Br. Zool.* I. *tab.* XXXVIII.

Found in all the quarters of the world. Flies fwiftly, ftrong, and direct. All the fpecies do not haunt rivers, nor prey on fifh.

2

IV

1

1. CAPREOWS JACKAMAR   2. YELLOW-CHEEKED-CREEPER

RED-HEADED KING-FISHER.

1. GREEN TODY. 2. BROWN TODY.

fiſh. 'Αλκυων of *Ariſtotle*, the *Halcyon* of *Pliny*, *Alcedo* of Linnæus, *Iſpida* of *Briſſon*, IV 471. The firſt gives us XV ſpecies.
Briſſon XXVI.

BILL, ſtrait ; on the lower mandible a ſmall angle.	XXIV. NUT-
NOSTRILS, ſmall, covered with feathers reflected over them.	HATCH.
TONGUE, ſhort, horny at the end, and jagged.	
TOES, three forward, one backward. The middle toe joined cloſely at the baſe to both the outmoſt. Back toe as large as the middle toe.	
Its manners, *vide Br. Zool.*	

Nuthatch *Br. Zool. fol. tab.* H. *Br. Zool.* I. *tab.* XXXVIII.   **Ex.**
Inhabits *Europe*, *Aſia*, *America*.
Sitta of LINNÆUS and *Briſſon* III. 588. he deſcribes V ſpecies,
   LINNÆUS II. *Ariſtotle*'s Ϛιττη not eaſily determinable.

BILL, thin, depreſſed, broad, baſe beſet with briſtles.	XXV. TODY.
NOSTRILS, ſmall.	
TONGUE.	
TOES, three forward, one backward, connected like thoſe of the Kingsfiſher.	

Green-ſparrow, EDW. CXXI.   **Ex.**

Inhabits the hot parts of *America*.
Todus of LINNÆUS and *Briſſon* IV. 528. who enumerate II ſpecies.
   The name firſt given it by Dr. *Brown*, I ſuppoſe, from *Todi*,
   ſmall birds.

G

BILL,

XXVI. BEE-EATER.
BILL, quadrangular, a little incurvated, sharp pointed.

NOSTRILS, small, placed near the base.

TONGUE, slender.

TOES, three forward, one backward: The three lower joints of the middle toe closely joined to those of the outmost.

Ex.
Indian Bee-eater, EDW. CLXXXIII.

Feeds on Bees, which it catches in its flight; from which the *English* name.

Inhabits *Southern Europe, Asia, Africa*, and *America*.

Merops of LINNÆUS, *Apiaster* of *Brisson*, IV. 532. The first has VII species, the last XIII.

XXVII. HOOPOE
BILL, long, slender, and bending.

NOSTRILS, small, placed near the base.

TONGUE, short, sagittal.

TOES, three forward, one backward; middle toe closely united at the base to the outmost.

Ex.
Hoopoe, *Br. Zool. fol. tab.* L. *Br. Zool.* I. *tab.* XXXIX.

Inhabits *Europe* and *Asia*.

Upupa of LINNÆUS. *Upupa* and *Promerops* of *Brisson*, II 456. 460. LINNÆUS has III species. *Brisson* I of the first, V of the last.

XXVIII. CREEPER.
BILL, very slender, weak, incurvated.

NOSTRILS, small.

TONGUE, not so long as the bill; hard, and sharp at the point.

TOES, three forward, one backward; large back toe, and long hooked claws.

Creeper,

INDIAN BEE EATER.

HONEYSUCKERS

Creeper, *Br. Zool. fol. tab.* K. *Br. Zool.* I. **XXXIX.** According
to its name, creeps up and down the trunks and branches of
trees, feeding on infects, their eggs and *larvæ*.

Inhabits *Europe, Afia, Africa,* and *America.*

Certhia of Linnæus and *Briffon,* III. 602. The firft has **XXV**
fpecies; the laft **XXXII.**

BILL, flender and weak; in fome ftrait, in others incurvated.
Nostrils, minute.

Tongue, very long, formed of two conjoined cylindric tubes;
miffile.

Toes, three forward, one backward.

Tail, confifts of ten feathers.

Long tailed red humming bird, Edw. **XXXII.** which is the fort
with crooked bills, called by *Briffon, Polytmi.*

White bellied humming bird, Edw. **XXXV.** or the kind with
ftrait bills: The *Mellifuga* of *Briffon.*

Feeds on the fweet juices of flowers, which it fucks out with its
tubular tongue, hanging in the air on its wings.

Inhabits *America,* efpecially the warm parts: A numerous genus.

Trochilus of Linnæus; *Polytmus* and *Mellifuga* of *Briffon.* Linnæus has **XXII** fpecies; *Briffon* **XVI** of the *Polytmus,* **XX** of
the *Mellifuga.* The old *Englifh* name was *Humming-bird;* which
I now change to Honey-Sucker.

ORDER

ORDER III.

GALLINACEOUS.

Heavy bodies, fhort wings, very convex; ftrong, arched, fhort bills: The upper mandible fhutting over the edges of the lower. The flefh delicate, and of excellent nutriment; ftrong legs; toes joined at the bafe, as far as the firft joint, by a ftrong membrane. Claws broad, formed for fcratching up the ground. More than twelve feathers in the tail.

Granivorous, feminivorous, infectivorous, fwift runners, of fhort flight; often polygamous, very prolific, lay their eggs on the bare ground. Sonorous, querulous, and pugnacious.

Or, with bills flightly convex; granivorous, feminivorous, infectivorous; long legs, naked above the knees: The genus that connects the land and the water-fowl. Agreeing with the cloven-footed water-fowl in the length and nakednefs of the legs, and the fewnefs of its eggs: Difagreeing in place, food, and form of bill, and number of feathers in the tail.

BILL, very convex, fhort, and ftrong.   XXX. COCK.

NOSTRILS, bodied in a flefhy fubftance.

TONGUE, cartilaginous, fharp, entire.

HEAD, adorned with a *Comb*, or elevated ferrated flefh.

SPURS on the legs.

TAIL, confifting of fourteen feathers; that of the male, fickle-
  fhaped.

To be found in every farm-yard.   Ex.

Its native country *India* and its ifles.

Domefticated every where.

Phafianus of LINNÆUS, who claffes it with the Pheafant, and has
  VI fpecies.   *Gallus* of *Briffon*, I. 165. who enumerates V, but
  they are only varieties.

BILL, convex, fhort and ftrong.   XXXI. TURKEY.

NOSTRILS, open, pointed at one end, lodged in a membrane.

TONGUE, floped on both fides towards the end, and pointed.

HEAD and NECK, covered with a naked tuberofe flefh, with a long
  flefhy appendage hanging from the bafe of the upper mandible.

TAIL, broad, confifts of eighteen feathers, extenfible.

Unknown to none.   Ex.

Native of *North America* only: Domefticated in moft countries.

Meleagris of LINNÆUS, and *Gallo-pavo* of *Briffon*, I. 158. LIN-
  NÆUS has III, *Briffon* II fpecies.

BILL,

XXXII. PIN-
TADO.

BILL, convex, ftrong, and fhort; at the bafe a carunculated cere,
in which the

NOSTRILS are lodged.

HEAD and NECK, naked, flightly befet with briftles.

A HORN, reflected and large, on the head.

LONG POINTED WATTLES, hanging from the cheeks.

TAIL, fhort, pointing downwards.

Ex.

Too common to need a reference.

Its native place *Africa*.

Numida of LINNÆUS, who has I. fpecies. *Meleagris* of *Briffon*,
I. 176. who has likewife I. He calls it in French, *La Peintade*,
a name I retain.

XXXIII. CU-
RASSO.

BILL, convex, ftrong, and thick, the bafe covered with a cere,
often mounted by a large nob.

NOSTRILS, fmall, lodged in the cere.

HEAD, fometimes adorned with a creft of feathers, curling forwards.

TAIL, large, ftrait.

Ex.

Curaffo, and Cufhew-bird, EDW. CCXCV.

Inhabits *South America*.

Crax of LINNÆUS and *Briffon*, I. 296. But the laft claffes them
with the Pheafant, and has VI fpecies, LINNÆUS III.

XXXIV. PEA-
COCK.

BILL, convex, ftrong, and fhort.

NOSTRILS, large.

HEAD, fmall, crefted.

SPURS on the legs.

TAIL,

Tail, very long, broad, expanfible, confifting of a double range
of feathers, adorned with rich ocellated fpots.

Common Peacock, frequent in moft parts. The Peacock Phea-   Ex.
fant, Edw. LXVII.
The native place *India, Japan,* and *China.*
Pavo of Linnæus, and *Phafianus* of *Briffon,* I. 281. who reckons
IV fpecies of Peacocks, Linnæus III.

Bill, convex, fhort, and ftrong.   XXXV. PHEA-
Nostrils, fmall.   SANT.
Tail, very long, cuneiform, bending downwaads.

Painted Pheafant, Edw. LXVIII.
Inhabits *Afia* and *South America.*
Phafianus of Linnæus and *Briffon* I. 262. who has (including
Peacocks and Curaffoas) XVI fpecies, Linnæus VI.

Bill, convex, ftrong, and fhort.   XXXVI. GROUS.
A naked fcarlet fkin above each Eye.
Nostrils, fmall, and hid in the feathers.
Tongue, pointed at the end.
Legs, ftrong, feathered to the toes; and fometimes to the nails.
The toes of thofe with naked feet pectinated on each fide.

Grous, *Br. Zool. fol. tab.* M. 3. *Br. Zool.* I. *tab.* XLIII.   Ex.
Inhabits the mountains or woods of *Europe,* northern and eaftern
Afia, and *North America.*

TETRAO

TETRAO *pedibus hirfutis* of LINNÆUS, who has IX fpecies. *Lagopus* of *Briffon*, I. 181. who has XII.

XXXVII. PAR-TRIDGE.

BILL, convex, ftrong, and fhort.

No naked fkin above the EYES.

NOSTRILS, covered above with a callous prominent rim.

LEGS, naked, tetradactylous. *Exception*, two fpecies of Quails.

TAIL, fhort.

Ex.

Partridge, *Br. Zool. fol. tab.* M. V.

Inhabits the cultivated parts of the world.

TETRAO *pedibus nudis* of LINNÆUS, who has XI fpecies. *Perdix* of *Briffon*, who has XXI.

XXXVIII. TRUMPETER.

BILL, fhort, upper mandible a little convex.

NOSTRILS, oblong, funk, and pervious.

TONGUE, cartilaginous, flat, torn, or fringed at the end.

LEGS, naked a little above the knees.

TOES, three before; one fmall behind, with a round protuberance beneath the hind toe, which is at a fmall diftance from the ground.

Ex.

Grus Pfophia, *Pallas fpicil. fafc.* IV. *tab.* I.

Inhabits *South America*; lives in the woods; feeds on the fruit that fall down. Does not perch. Makes a ftrong noife with its mouth, which it anfwers by a different noife from its belly, as if it came from the anus. Lays many eggs.

Pfophia of LINNÆUS, from ψοφεω *ftrepitum edo*. *Perdix* of *Briffon*, I. 227 only I. fpecies. A beautiful fpecimen in the LEVERIAN *Mufeüm*.

BILL

Bill, a little convex.

Nostrils, open, oblong.

Tongue, floping on each fide near the end, and pointed.

Legs, long, and naked above the knees.

Toes, only three; no back toe.

XXXIX. BUS-
TARD.

Bustard, *Br. Zool. fol. tab.* IV. *Br. Zool.* I. *tab.* XLIV.

Inhabits *Europe, Afia, Africa,* and *New Holland.*

Otis of Linnæus and *Briffon,* V. 18. One has IV. the other III
fpecies. *De Buffon. Pliny* tells us, that *Otis* was the *Greek*
name, that the *Spanifh* was *Sarda.*

Ex.

H ORDER

ORDER IV.

C O L U M B I N E.

Bill, weak, flender, ftrait at the bafe, with a foft protuberant fub-ftance, in which the noftrils are lodged. Tongue, entire: Legs, fhort, and red: Toes, divided to the origin. Swift and diftant flight, walking pace. Plaintive note, or *cooing*, peculiar to the order. The male inflates or fwells up its breaft in court-fhip. Female, lays but two eggs at a time. Male and female fit alternately; and feed their young, ejecting the meat out of their ftomachs into the mouths of the neftlings. Granivorous, feminivorous. The neft fimple, in trees, or holes of rocks, or walls.

XL. PIGEON.   There is only one genus of this order; it is therefore needlefs to repeat the characters.

Ex.   A well known bird.
Inhabits all the Continents.
Columba of Linnæus and *Briffon*, I. 67. Linnæus has XL fpe-cies, *Briffon* XLIV.

ORDER V.

PASSERINE.

BODIES, from the fize of a Thrufh, to that of the golden-crefted
Wren. The enliveners of the woods and fields; fprightly, and
much in motion; their nefts very artificial; monogamous, bac-
civorous, granivorous, feminivorous, infectivorous; their ufual
pace, hopping; of a few, running. Short flyers, except on
their migrations only. All have three Toes before, one behind.

BILL, ftrait, depreffed.   XLI. STARE.
NOSTRILS, guarded above by a prominent rim.
TONGUE, hard and cloven.
TOES, the middle joined to the outmoft as far as the firft joint.

Stare, *Br. Zool. fol tab.* P. II. *Br. Zool.* I. *tab.* XLVI.   Ex.
Sturnus of LINNÆUS and *Briffon* II. The firft has V fpecies, the
laft four.

BILL, ftrait, obtufely carinated at top, bending a little at the XLII. THRUSH.
point, and flightly notched near the end of the upper mandible.
NOSTRILS, oval and naked.
TONGUE, flightly jagged at the end.

H 2

TOES,

Toes, the middle joined to the outmoft as far as the firft joint; back toe very large.

Ex. Fieldfare, *Br. Zool. fol.* P. II.
Blackbirds, *Br. Zool.* I. *tab.* XLVII.
Turdus of Linnæus and *Briffon* II.
Linnæus has XXVIII fpecies, *Briffon* LXIV.

XLIII. CHAT-TERER. Bill, ftrait, a little convex above, and bending towards the point; near the end of the upper mandible, a fmall notch on each fide.
Nostrils, hid in briftles.
Middle Toe, clofely connected at the bafe to the ontmoft.

Ex. The Pompadour, Edw. CCCXLI.
Ampelis of Linnæus (from αμπελος, a vine); becaufe the *Bohemian* Chatterer, the bird he places at the head of this genus, feeds fometimes on grapes. He reckons VII fpecies. The *Cotinga* of *Briffon* II. 339. an *American* name. He has X fpecies. Inhabits *Europe* and *America*.

XLIV. COLY. Bill, convex above, ftrait beneath; very fhort and thick.
Nostrils, fmall, placed at the bafe, and hid by the feathers.
Tongue, not the length of the bill, laciniated at the end.
Toes, divided to their origin.

Ex. *Le Coliou, Briffon* III. *part* I. *tab.* XVI. *fig.* 2.
Inhabits *Africa*.
Linnæus includes this among his *Loxiæ*. *Briffon* III. *part* I. 304. calls it *Colius*.

Bill,

PURPLE CHATTERER.

Bill, ftrong, and convex above and below, very thick at the bafe.
Nostrils, fmall and round.
Tongue, as if cut off at the end.

Grofbeak, *Br. Zool. fol. tab.* U.
Pine Grofbeak, *Br. Zool.* I. *tab.* XLIX.
Inhabits every Continent.
Loxia of Linnæus including the *Coccothrauftes* of *Briffon*, III.
part I. 219. the *Colius* 304. the *Pyrrhula* 308. and *Loxia* or
Crofs-bill 329. Linnæus has XLVII fpecies; *Briffon* in all
XXXI. *Loxia* is the proper name of the Crofs-bill, from
λοξος, oblique.

Bill, ftrong, and conic, the fides of each mandible bending inwards; in the roof of the upper mandible, a hard knob, of
ufe to break and comminute hard feeds.

Bunting, *Br. Zool. fol. tab.* W.
Inhabits *Europe, Afia,* and *America.*
Emberiza of Linnæus and *Briffon* III. part I. 257. The firft has
XXIV fpecies, the laft XV. The name is derived from *Em-
britz,* or *Emmeritz,* its *German* name. *Vide Gefner, av.* 653.

Bill, conoid, a little inclining towards the point, upper
mandible flightly ridged, and notched near the end.

Red-breafted Blackbird, Edw. CCLXVII. and greater Bulfinch,
LXXXII.
Inhabits *North* and *South America;* moft numerous in the latter.

Tanagra

Tanagra of Linnæus, and *Tangara* of *Briſſon* III. *part* I. 3.
Linnæus reckons XXIV ſpecies, *Briſſon* XXX.
The name *Tangara* is *Braſilian.*

XLVIII. FINCH.　Bill, perfectly conic, ſlender towards the end, and very ſharp
　　　　　　　　pointed.

　　　　　　　Goldfinch, *Br. Zool. fol. tab.* V.
　　　　　　　Sparrows, *Br. Zool.* I. *tab.* XLI.
　　　　　　　Inhabits all the quarters of the world.
　　　　　　　Fringilla of Linnæus, who enumerates XXXIX ſpecies.　*Paſſer*
　　　　　　　of *Briſſon* III. part I. 71. who has LXVII ſpecies.

XLIX. FLY-　　Bill, flatted at the baſe, almoſt triangular, notched at the end of
　CATCHER.　　the upper mandible, and beſet with briſtles.
　　　　　　　Toes, divided as far as their origin.

　　Ex.　　　Flycatcher, *Br. Zool. fol. tab.* P. II.
　　　　　　　Inhabits all the quarters of the world.
　　　　　　　Muſcicapa of Linnæus and *Briſſon*, II.　The firſt has XXI ſpe-
　　　　　　　cies ; the laſt XXXVIII.

L. LARK.　　Bill, ſtrait, ſlender, bending a little towards the end, ſharp
　　　　　　　pointed.
　　　　　　　Nostrils, covered with feathers and briſtles.
　　　　　　　Tongue, cloven at the end.

　　　　　　　　　　　10

　　　　　　　　　　　　　　　　　　　　　　　　　　Toes,

TOES, divided to the origin; claw of the back toe very long, and either ftrait, or very little bent.

Larks, *Br. Zool. fol. tab.* S. *Br. Zool.* I. *tab.* LV.   Ex.
Inhabits *Europe, Afia, Africa,* and *America.*
Alauda of LINNÆUS and *Briſſon* III. part II. 33 . LINNÆUS has
   XI fpecies, *Briſſon* XII.

BILL, weak and flender.   LI. WAGTAIL.
TONGUE, lacerated at the end.
LEGS, flender.
Frequent the fides of brooks; their tails much in motion; their
   pace running; feldom perch; their neſt on the ground.

Wagtails, *Br. Zool.* I. *tab.* LV.   Ex.
After the example of *Scopoli,* I feparate thefe, the genuine *Motacillæ,* from the other foft-bill'd fmall birds, which he ftiles
   Sylviæ. They are included among the *Motacillæ* of LINNÆUS,
   and *Ficedulæ* of *Briſſon* III. part II. 369.

BILL, flender and weak.   LII. WARBLERS.
NOSTRILS, fmall, funk.
TONGUE, cloven.
FEET, the exterior toe joined at the under part of the laſt joint
   to the middle toe.

Red-breaſt, *Br. Zool. fol. tab.* S.   Ex.

Inhabits all parts of the world, except the *Arctic :* The moft melodious of the fmaller *genera :* Infectivorous, feminivorous, de-
light in woods and bufhes.   Their pace hopping.   *Motacilla*
of LINNÆUS, *Ficedula* of *Briffon* III. part II. 369.   LINNÆUS
has XLIX fpecies, *Briffon* LXXIII.

LIII. MANAKIN.   BILL, fhort, ftrong, and hard, flightly incurvated.

NOSTRILS, naked.

TONGUE.

TOES, the middle clofely united with the outmoft as far as the
third joint.

TAIL, fhort.

Ex   Manakins, EDW. CCLXI.

Inhabits *South America* only.

Pipra of LINNÆUS, and *Manacus* of *Briffon* IV. 442.   LINNÆUS
enumerates XIII fpecies, *Briffon* XIII.

Pipra, a πιπρα, a certain bird, mentioned by *Ariftotle, hift. an. lib.*
IX. C. I.   *Manacus* from the *Dutch, Manakin,* the name they
bear in *Surinam.*

LIV. TITMOUSE   BILL, ftrait, a little compreffed, ftrong, hard, and fharp pointed.

NOSTRILS, round, and covered with briftles reflected over them.

TONGUE, as if cut off at the end, and terminated by three or four
briftles.

TOES, divided to their origin; back toe very large and ftrong.

Titmice,

2

1

1. CRESTED MANAKIN. 2. GOLDEN HEADED M.

Titmice, *Br. Zool. fol. tab.* W. *Br. Zool.* I. *tab.* LVII. a reftlefs   Ex.
fliting race; moft prolific; infectivorous, germinivorous, pugnacious.

Inhabit *Europe* and *America.*

Parus of LINNÆUS and *Briffon* III. part II. 539. LINNÆUS has
XIV. *Briffon* XVIII fpecies. *Parus,* from *Pario,* becaufe it
lays many eggs.

BILL, fhort, broad at the bafe, fmall at the point, and a little   LV. SWALLOW.
bending.

NOSTRILS, open.

TONGUE, fhort, broad, and cloven.

LEGS, fhort.

TAIL, forked; Wings, long.

Inhabits the univerfe, even as far as *Hudfon's Bay.*

Swallow, *Br. Zool. fol. tab.* Q. *Br. Zool.* I. *tab.* LVIII. Swift,   Ex.
much on wing, infectivorous, migratory or torpid during winter, twittering, forerunners of fummer.

Hirundo of LINNÆUS and *Briffon* II. 485. LINNÆUS has XII
fpecies, *Briffon* XVII.

BILL, very fhort, hooked at the end, and very flightly notched   LVI. GOAT-
near the point.   SUCKER.

NOSTRILS, tubular, and a little prominent.

MOUTH, vaftly wide: On the edges of the upper part, between
the bill and the eyes, feven ftiff briftles.

TONGUE, fmall, entire at the end.

I   LEGS,

Legs, fhort, feathered before as low as the toes.

Toes, joined by a ftrong membrane as far as the firft joint. Claw of the middle toe broad-edged and ferrated.

Tail, confifts of ten feathers, and is not forked.

Inhabits *Europe*, *Afia*, and *America*.

Flies by night; infectivorous, fonorous, migratory. Has much of the nature of the Swallow.

Ex. Goatfucker, *Br. Zool. fol. tab.* R. *Br. Zool.* I. *tab.* LIX. *Caprimulgus* of Linnæus, *Caprimulgus* and *Tette-chévre* of *Briffon* II. 470. Linnæus has only II fpecies, *Briffon* VI. *Caprimulgus* and Αιγοθηλης of the Ancients, from a vulgar notion that they fucked the teats of Goats.

ORDER VI.

STRUTHIOUS.

Very great and heavy Bodies. Wings, imperfect; very fmall, and ufelefs for flight, but affiftant in running. Flefh coarfe, and hard of digeftion.

Struthious is a new coined word to exprefs this order; for thefe birds could not be reduced to any of the Linnæan divifions.

Bill, large, bending inward in the middle of the upper man-   LVII. DODO.
dible, marked with two oblique ribs, and much hooked at
the end.
Nostrils, placed obliquely near the edge, in the middle of the
bill.
Legs, fhort, thick, feathered a little below the knees.
Toes, three forward, one backward.

Dodo, Edw. CCXCIV.   Ex.
Inhabits the ifles of *France* and *Bourbon.*
Didus of Linnæus, and *Raphus* of *Briffon,* V. 14. only I. fpecies.

I 2   Bill,

LVIII. OSTRICH　Bill, fmall, floping, a little depreffed.

　　　　　　　　Small Wings, unfit for flight.

　　　　　　　　Legs, long, ftrong, naked above the knees.

　　Ex.　　　　Oftrich and Caffowary, *Wil. Orn.* tab. XXV.

　　　　　　　　Inhabits *Afia, Africa,* and the lower parts of *South America.*

　　　　　　　　Struthio of Linnæus and *Briffon* V. 3. III fpecies.

D I V.

DIV. II. WATER-FOWL.

ORDER VII. With CLOVEN FEET.

VIII. With PINNATED FEET.

IX. With WEBBED FEET.

Moſt migratory, ſhifting from climate to climate, from place to place, in order to lay their eggs, and bring up their young in full ſecurity: the thinly inhabited north their principal breeding place; returning at ſtated periods, and, in general, yielding to mankind delicious and wholeſome nutriment. All the *Cloven-footed*, or mere Waders, lay their eggs on the ground. Thoſe with *pinnated feet* form large neſts, either in the water, or near it. From the firſt, we muſt except the *Heron* and the *Night-Heron** *, which build in trees.

All the Web-footed fowl either lay their eggs on the ground, or on the ſhelves of lofty cliffs; and none perch, except the Corvorant, Shugg, and one or two ſpecies of Ducks.

* Night Raven, *Raii Syn. av.* 99.

All

All the Cloven-footed Water-fowl have long necks and long legs, naked above the knees, for the convenience of wading in waters in fearch of their prey. Thofe that prey on fifh have ftrong bills. Thofe that fearch for minute infects, or worms that lurk in mud, have flender weak bills, and olfactory nerves of moft exquifite fenfe; for their food is out of fight.

As the name implies, their toes are divided, fome to their origin; others have, between the middle toe and outmoft toe, a fmall membrane as far as the firft joint. Others have both the exterior toes connected to the middlemoft in the fame manner; and, in a few, thofe webs reach as far as the fecond joint; and fuch are called *Semipalmati*.

Of the Web-footed fowl, the *Flamingo*, the *Avofetta*, and *Courier*, partake of the nature of both the Cloven and Web-footed orders; having webbed feet, long legs, naked above the knees, and long necks. The other Web-footed Water-fowl being very much on the element, have fhort legs, placed far behind, and long necks; and, when on land (by reafon of the fituation of their legs) an aukward waddling gate.

The make of the Cloven-footed Water-fowl is light, both as to fkin and bones; that of the Web-footed, ftrong.

ORDER

ORDER VII.

CLOVEN-FOOTED.

BILL, long, broad, flat, and thin, the end widening into a circular form like a spoon.

LIX. SPOON-BILL.

NOSTRILS, small, placed near the base.

TONGUE, small and pointed.

FEET, semipalmated.

Spoon-bill, *Wil. orn. tab.* 52. *Br. Zool.* II. *App.*

Ex.

Inhabits *Europe, South America,* and the *Philippine Islands* *.

Breeds in high trees; feeds on fish, and water-plants; can swim.

Platalea of LINNÆUS, and *Platea* of *Brisson* V. 351. Each have III species.

BILL, bending down at the point, with a horn, or with a tuft of feathers erect near the base of the bill.

LX. SCREAMER.

NOSTRILS, oval.

TONGUE.

TOES, divided almost to their origin, with a very small membrane between the bottoms of each.

* Voy. de *Sonnerat.* 89.

2

Anhima

Ex.

Anhima *Marcgrave* 215.

Inhabits *South America.*

Palamedea of Linnæus, *Anhima* and *Cariama* of *Briffon* V. 518. I call it Screamer, from the violent noife it makes. Only two fpecies.

LXI. JABIRU.

BILL, long, and large, both mandibles bending upwards; the upper, triangular.

NOSTRILS, fmall.

No TONGUE? *Marcgrave.*

TOES, divided.

Ex.

Jabiru guacu *Marcgrave* 200. 201.

Inhabits *South America.*

Myƈteria of Linnæus, from Μυχτηρ, a fnout. *Ciconia* of *Briffon* V. 371. Only one fpecies.

LXII. BOAT-BILL.

BILL, broad, flat, with a keel along the middle, like a boat reverfed.

NOSTRILS, fmall, lodged in a furrow.

TONGUE.

TOES, divided.

Ex.

Tamatia *Marcgrave* 208. 209. *Brown's Zool.* 92. *tab.* XXXVI.

Inhabits *South America.*

Cancroma of Linnæus, from their feeding on Crabs, who has II fpecies; the *Cochlearius* of *Briffon* V. 206. who has the fame number.

BILL,

BILL, long, ſtrong, ſharp pointed.

NOSTRILS, linear.

TONGUE, pointed.

TOES, connected as far as the firſt joint by a membrane ; back toe large.

LXIII. HERON.

Creſted Heron, *Br. Zool. fol. tab.* A.

Female Heron, *Br. Zool.* II. *tab.* LXI.

Inhabits every continent.

Ardea of LINNÆUS. *Ardea, Ciconia,* and *Balearica* of *Briſſon* V. 361. 391. 511. LINNÆUS has XXVI ſpecies, *Briſſon* LX.

Ex.

BILL, ſtrong, thick, ſtrait, compreſſed, the upper mandible com-poſed of ſeveral pieces.

LXIV. UMBRE.

Brown's Zool. 90. *tab.* XXXV.

Inhabits *Senegal* and the South of *Africa.*

Scopus of *Briſſon,* who has a ſingle ſpecies. He calls it *Scopus,* from σκια, a ſhade ; and *Ombrette* from the general deep brown of its plumage.

Ex.

BILL, long, thick at the baſe, wholly incurvated. EYES, lodged in the baſe.

FACE, naked.

NOSTRILS, linear.

TONGUE, ſhort and broad.

TOES, connected at the baſe by a membrane.

LXV. IBIS.

Red

Ex.   Red Curlew *Catesby Carol.* I. LXXXIV. White-headed *Ibis, Ind. Zool. tab.* X.

Inhabits *Europe, Asia,* and *America.*

Tantalus of LINNÆUS, *Numenius* of *Brisson* V. 311. LINNÆUS has VII species. *Brisson* mixes them with the genuine Curlews, and has in all XIV.

LXVI. CURLEW.   BILL, long, slender, incurvated.

FACE, covered with feathers.

NOSTRILS, linear, longitudinal, near the base.

TONGUE, short, and sharp pointed.

TOES, connected as far as the first joint by a strong membrane.

Ex.   Curlew *Br. Zool.* II. *tab.* LXIII.

Inhabits *Europe, America,* the *Philippine Isles,* and *New Holland.*

Scolopax of LINNÆUS, *Numenius* of *Brisson* V. 311. LINNÆUS has IV species of genuine Curlews.

LXVII. SNIPE.   BILL, two inches long and upwards; slender, strait, and weak.

NOSTRILS, linear, lodged in a furrow.

TONGUE, pointed, slender.

TOES, divided, or very slightly connected; back toe very small.

Ex.   Woodcock, *Br. Zool.* II. *tab.* LXV.

Inhabits *Europe, Asia,* and *America.*

Scolopax of LINNÆUS, *Limosa* and *Scolopax* of *Brisson* V. 261. 292.

LINNÆUS reckons XIV species, exclusive of the Curlews. *Brisson* XIII. Woodcock being the name of a species inha-

9   biting

DWARF CURLEW.

LITTLE SANDPIPER.

biting woods, I change it to the more comprehenſive one of *Snipe,* which ſignifies *a long bill.*

BILL, ſtrait, ſlender, and not an inch and a half long.

NOSTRILS, ſmall.

TONGUE, ſlender.

TOES, divided; generally the two outmoſt connected at bottom by a ſmall membrane.

LXVIII. SAND-
PIPER.

Purr *Br. Zool.* II. *tab.* LXXI.

Inhabits all the quarters of the world; but in greateſt plenty in the *Northern.*

Tringa of LINNÆUS; *Vanellus, Arenaria, Glareola,* and *Tringa* of *Briſſon* V. 94. 132. 141. 177. including XXXV ſpecies.

Ex.

BILL, ſtrait, as ſhort as the head.

NOSTRILS, linear.

TONGUE.

TOES. Wants the back toe.

LXIX. PLOVER.

Dotterel *Br. Zool.* II. *tab.* LXXIII.

Charadrius of LINNÆUS, *Pluvialis* of *Briſſon* V. 43. and *Himan-topus* and *Oſtralega* 33. and 38. LINNÆUS has XII. *Ch. Briſſon* XV. *Pl.* II. *Himan.*

Ex.

BILL, long, compreſſed, the end cuneated.

NOSTRILS, linear.

TONGUE, ſcarce a third the length of the bill.

TOES, only three; the middle joined to the exterior by a ſtrong membrane.

LXX. OYSTER-
CATCHER.

K 2

Sea-

Ex.   Sea-Pie, *Br. Zool. fol. tab.* D. 2. *Br. Zool.* II. *tab.* LXXIV.
Inhabits *Europe*, *North America*, and the eaſtern coaſt of *New Holland*. The bill calculated to raiſe limpets, oyſters, and other ſhells from the rocks.
Hæmatopus of LINNÆUS, *Oſtralega* and *L'Huitrier* of *Briſſon* V. 38. Only one ſpecies.

LXXI. JACANA.   BILL, ſlender, ſharp pointed; thickeſt towards the end; the baſe carunculated.
NOSTRILS, ſhort, ſub-ovated, placed in the middle of the bill.
TONGUE.
WINGS, armed on the front joint with a ſharp, ſhort ſpur.
TOES, four on each foot, armed with very long and ſtrait ſharp pointed claws.

Ex.   Spur-winged Water Hen.   EDW. CCCLVII.
Parra of LINNÆUS, *Jacana* of *Briſſon* V. 122. LINNÆUS has has only III. genuine ſpecies, *Briſſon* V. I retain the *Braſilian* name *Jacana*. Is not the *Impios* PARRÆ *recinentis omen* of *Horace*, which was probably ſome ſmall bird. *Vide Pliny*, lib. X. c. 33.

LXXII. PRATIN-   BILL, ſhort, ſtrong, ſtrait, hooked at the end.
COLE.   NOSTRILS, near the baſe, linear, oblique.
TONGUE.
TOES, long, ſlender, baſe of each connected by a very ſmall membrane.
TAIL, forked; twelve feathers.

Pratincola,

Pratincola, *Kramer Auftr.* 382.   Ex.

Inhabits *Southern Europe.*

Pratincola, or inhabitant of meadows, a name given it by Dr.
Kramer, and adopted by me; placed by Linnæus with the
Hirundo, by *Briffon* among his *Glareolæ.*

Bill, flender, a little compreffed, and flightly incurvated.   LXXIII. RAIL.

Nostrils, fmall.

Tongue, rough at the end.

Body, much compreffed.

Tail, very fhort.

Water-Rail, *Br. Zool.* II. *tab.* LXXV.   Ex.

Inhabits *Europe, Afia,* and *America.*

Rallus of Linnæus, who places it among others very different,
fuch as the Land-Rail, &c. *Briffon* calls the genus *Rallus,* but
mixes with it others of another genus.

Bill, ftrong, thick, a little convex: upper of the upper mandible   LXXIV.
lodged in a corneous fheath. Sometimes elevated and open in   SHEATH-BILL.
front: at other times clofely applied to the bill; reaching beyond the edges of the mandible.

Nostrils, fmall: juft appearing out of the fheath.

Orbits, naked, granulated.

Wings, armed at the fecond flexure with a hard knob.

Legs and Toes, thick, gallinaceous. Toes edged with a
thick membrane. The middle toe connected to the next

by

by a web, as far as the first joint. CLAWS, blunt, guttered below.

TONGUE, fagittal, blunted at the point.

A new genus. Frequents watry places in *New Zeland* and *Statenland.*

LXXV. GALLI- BILL, thick at the bafe, floping to the point; the upper mandible
NULE.   reaching far up the forehead, and not corneous.

BODY, compreffed.

WINGS, fhort and concave.

TOES, long, divided to their origin.

TAIL, fhort.

EX.   Water-Hen, *Br. Zool. fol. tab.* L. I. *Br. Zool.* II. *tab.* LXXVII.
              Inhabits *Europe, Afia,* and *America.*

Fulica of LINNÆUS, *Gallinula* VI. and *Porphyrio* V. ₅22: of *Briffon,*
              who has III fpecies of the firft, and X of the laft. In Bill and
              Legs, the *Land-Rail* agrees with this genus; but, with us,
              differs in its manners, by refiding in dry places. But, as it mi-
              grates at approach of winter, it may, in warmer climates, dur-
              ing the feafon, inhabit fenny tracts, to which the form of its
              legs are adapted.

ORDER

O R D E R   VIII.

With P I N N A T E D   F E E T.

BILL, ſtrait, ſlender.　　　　　　　　　　　　　LXXVI. PHALA-
　　　　　　　　　　　　　　　　　　　　　　　　ROPE.
NOSTRILS, minute.

TONGUE.

BODY and LEGS, in every reſpect formed like the Sand-piper.

TOES, furniſhed with ſcalloped membranes.

Scallop-toe'd Sand-piper, *Br. Zool. fol. tab.* E. *Br. Zool.* II. *tab.*　　Ex.
　LXXVI.

Its manners, &c. unknown.

Inhabits *Europe* and *North America.*

LINNÆUS places it among the *Tringæ*; *Briſſon* very judiciouſly
　forms a new genus, under the name of *Phalaropus,* from the
　ſcallops on the toes, like the φαλαρις, or Coot.

BILL, ſhort, ſtrong, thick at the baſe, ſloping to the end; the LXXVII. COOT.
　baſe of the upper mandible riſing far up the forehead; both
　mandibles of equal length.

　　　　　　　　　　　　　　　　　　　　　NOSTRILS,

NOSTRILS, incline to oval, narrow, fhort.

TONGUE.

BODY, compreffed. WINGS, fhort.

TOES, long, furnifhed with broad fcalloped membranes.

TAIL, fhort.

Ex.

Coot, *Br. Zool. fol. tab.* F. *Br. Zool.* II. *tab.* LXXVII.

Inhabits *Europe, Afia,* and *Africa.*

Continues much on the water, makes a large neft of water-plants, lays fix or feven eggs. In winter, ofttimes are feen in great flocks on arms of the fea.

Fulica of LINNÆUS, and *Briffon* VI. 23. LINNÆUS has only II fpecies, for he mixes other birds with them. *Briffon* has II.

LXXVIII.
GREBE.

BILL, ftrong, flender, fharp pointed.

NOSTRILS, linear.

TONGUE, flightly cloven at the end.

BODY, depreffed. FEATHERS, thick-fet, compact, and very fmooth and gloffy.

TAIL, none. WINGS, fhort.

LEGS, placed very far behind, very thin, or much compreffed; doubly ferrated behind.

TOES, furnifhed on each fide with a broad, plain membrane.

Ex.

GREBE, *Br. Zool. fol. tab.* K. *Br. Zool.* II. *tab.* LXXVIII.

Congenerous birds, found in moft countries; north as high as *Hudfon's-Bay,* and fouth as far as *lat.* 48. 30. and *long.* 58. 7.

eaft.

eaſt *. Linnæus mixes his birds of this genus with web-footed birds, ſuch as *Divers* and *Guillemots*, by the general name of *Colymbi*. *Briſſon* VI. 33. very judiciouſly ſeparates them, and has under the ſame name XI ſpecies.

* *Cook's Voy.* i. 48.   *Forſter's* i. 115.

L   ORDER

ORDER IX.

WEB-FOOTED.

With LONG LEGS.

LXXIX. AVO-
SET.

BILL, long, flender, very thin, and bending confiderably upwards.

NOSTRILS, narrow and pervious.

TONGUE, fhort.

FEET, palmated; the webs deeply femilunated between each toe; back toe very fmall.

Ex.

Avofetta, *Br. Zool. fol. tab.* G. *Br. Zool.* II. *tab.* LXXX.

Inhabits *Europe, North America,* and the weftern coaft of *New Holland* *.

Recurviroftra of LINNÆUS, *Avofetta* of *Briffon* VI. 537. Two fpecies.

LXXX. COU-
RIER.

BILL, fhort, ftrait.

NOSTRILS.

TONGUE.

LEGS, long. THIGHS, fhort. FEET, palmated; has a back toe.

* *Dampier,* iii. 85.

5

Trochilus,

Trochilus, *vulgo* Corrira *Aldr. av.* III. 288. *Wil. orn. tab.* LX.   Ex.
Inhabits *Italy.*

Corrira of *Briſſon* VI. 542. Only one ſpecies, and that probably
 never obferved fince the days of *Aldrovandus,* who is the only
 writer who feems to have feen it.

BILL, thick, large, bending in the middle, forming a ſharp angle, LXXXI. FLAM-
 the higher part of the upper mandible carinated; the lower,   MANT.
 compreſſed. The edges of the upper mandible ſharply denti-
 culated; of the lower, tranſverſely fulcated.

NOSTRILS, covered above with a thin plate, pervious, linearly
 longitudinal.

TONGUE, cartilaginous, and pointed at the end; the middle muf-
 cular, bafe glandular, on the upper part aculeated.

NECK, very long.

LEGS and THIGHS, of a great length.

FEET, webbed; the webs extend as far as the claws, but are
 deeply femilunated.

BACK TOE, very fmall.

Flamingo, *Cateſby Carol.* I. LXXIII.   Ex.
Inhabits *South America, Africa,* and rarely the *South* of *Europe.*
Phœnicopterus of LINNÆUS. Only one ſpecies.

With

With S H O R T L E G S.

LXXXII. ALBA-TROSS. BILL, ſtrong, bending in the middle, and hooked at the end of the upper mandible. That of the lower mandible abrupt, and the lower part inclining downwards.

NOSTRILS, opening forward, and covered with a large convex guard.

TONGUE.

TOES, no back toe.

Ex. Albatroſs, EDW. LXXXVIII.

Inhabits the iſlands and ſeas within the Tropics, and as far ſouth as *lat.* 67. 15. *long.* 39. 35. eaſt *. Two or three ſpecies have been added to this genus.

Diomedea exulans of LINNÆUS, *Albatroſs* of *Briſſon* VI. 127.

LXXXIII. AUK. BILL, ſtrong, thick, convex, compreſſed.

NOSTRILS, linear, placed near the edge of the mandible.

TONGUE, almoſt as long as the bill.

TOES, no back toe.

Ex. Puffin, *Br. Zool. fol. tab.* H.

Great Auk, *Br. Zool.* II. *tab.* LXXXI.

Inhabits the northern parts of *Europe, Aſia,* and *America*; gregarious in general; lays only one egg; theſe, the *Grebe, Divers,* and *Pinguins,* while on land, ſeldom move much.

* *Cook's* Voy. i 43.

Alca

Alca of LINNÆUS, *Fratercula* and *Alca* of *Briſſon* VI. 81. 85.
LINNÆUS has V. ſpecies, *Briſſon* IV. viz. I. of the *Fr.* or *Puffin*,
III of the *Alca*, or *Auk*.

BILL, ſlender, ſtrong, pointed, the upper mandible ſlightly bend-   LXXXIV GUIL-
ing towards the end; baſe covered with ſhort ſoft feathers.   LEMOT.
NOSTRILS, lodged in a hollow near the baſe.
TONGUE, ſlender, almoſt the length of the bill.
TOES, no back toe.

Guillemot, *Br. Zool. fol. tab.* H. 3.   Ex.
Leſſer Guillemot, *Br. Zool.* II. *tab.* LXXXII.
Inhabits the ſame places with the former, and lays only one egg.
LINNÆUS claſſes this genus with the *Colymbi.* The *Uria* of *Briſſon*
VI. 70 who has IV. ſpecies.

BILL, ſtrong, ſtrait, pointed, upper mandible the longeſt; edges LXXXV.DIVER.
of each bending inwards.
NOSTRILS, linear; the upper part divided by a ſmall cutaneous
appendage.
TONGUE, long and pointed, ſerrated at each ſide near the baſe.
LEGS, very thin and flat.
TOES, the exterior the longeſt; the back toe ſmall, joined to the
interior by a ſmall membrane.
TAIL, ſhort; conſiſts of twenty feathers.

Great Northern Diver, *Br. Zool. fol. tab.* K. 2.   Ex.
Imber, *Br. Zool. vol.* II. *tab.* LXXXIV.
Inhabits the north of *Europe, Aſia,* and *America.*

Colymbus

Colymbus of LINNÆUS, and *Mergus* of *Briſſon* VI. 104: The laſt
has VI ſpecies. LINNÆUS mixes with this genus *Grebes* and
Guillemots.

LXXXVI. SKIM- BILL, greatly compreſſed; lower mandible much longer than the
   MER.   upper.

NOSTRILS, linear and pervious.

TONGUE.

TOES, a ſmall back toe.

TAIL, a little forked.

  Ex.   Cutwater, *Cateſby Carol*. I. *tab*. XC.

Inhabits *India* and *America*.

Rynchops of LINNÆUS. *Rygchopſalia* of *Briſſon* VI. 223. I.
   ſpecies.

Rynchops, from ρυγχος, a bill, and κοπτειν, to cut; the upper mandible being as if cut. I call it *Skimmer*, from the manner of
   its collecting its food with the lower mandible, as it flies along
   the ſurface of the water.

LXXXVII. TERN BILL, ſtrait, ſlender, pointed.

NOSTRILS, linear.

TONGUE, ſlender and ſharp.

WINGS, very long.

TOES, a ſmall back toe.

TAIL, forked.

  Ex.   Great Tern, *Br. Zool. fol. tab*. L *.

Little Tern, *Br. Zool*. II. *tab*. XC.

                                            Clamorous,

Clamorous, much on the wing, gregarious, lays four eggs on the ground.

Inhabits *Europe*, and *America*, *North* and *South*.

Sterna of LINNÆUS, and *Briffon* VI. 202. LINNÆUS has VII fpecies, *Briffon* the fame.

BILL, ftrong, ftrait, but bending down at the point; on the under part of the lower mandible an angular prominency.

NOSTRILS, oblong and narrow, placed in the middle of the bill.

TONGUE, a little cloven.

BODY, light; WINGS, long.

LEGS, fmall, and naked above the knees; back toe fmall.

LXXXVIII.
GULL.

Great Black and White Gull, *Br. Zool. fol. tab.* L.

Winter Mew, *Br. Zool.* II. *tab.* LXXXVI.

An univerfal genus; much on the wing, clamorous, hungry, pifcivorous, carnivorous, gregarious; lays four or five eggs, generally on high cliffs.

Larus of LINNÆUS, and *Larus* and *Stercoraria* of *Briffon* VI. 153. The firft has XI fpecies, the laft XVIII.

Ex.

BILL, ftrait; but hooked at the end.

NOSTRILS, cylindric, tubular.

TONGUE.

LEGS, naked above the knees.

BACK TOE, none; inftead, a fharp fpur pointing downwards.

LXXXIX.
PETREL.

Fulmar, *Br. Zool. fol. tab.* M. 2.

Stormy Petrel, *Br. Zool.* IV. *tab.* XCI.

Ex.

Inhabits

Inhabits all parts of the ocean; but the fpecies moft numerous in the high fouthern latitudes, as the Auks are in the northern. Many have the faculty of fpurting an oily liquid out of their ftomachs.

Procellaria of LINNÆUS, *Procellaria* and *Puffinus* of *Briffon*. LINNÆUS has VI fpecies, *Briffon* VII.

XC. MERGAN- SER. BILL, flender, a little depreffed, furnifhed at the end with a crooked nail. Edges of each mandible very fharply ferrated.

NOSTRILS, near the middle of the mandible, fmall and fub-ovated.
TONGUE, flender.
FEET, the exterior toe longer than the middle.

Ex. Goofander, *Br. Zool. fol. tab.* N *.
Red-breafted Goofander, *Br. Zool.* II. XCIII.
Great divers, feed on fifh.
Mergus of LINNÆUS, and *Merganfer* of *Briffon*, VI. 230. LINNÆUS has VI fpecies, *Briffon* VII. The name, Merganfer; or Diving-Goofe.
Inhabits the *North* of *Europe*, and *North America*.

XCI. DUCK. BILL, ftrong, broad, flat, or depreffed; and commonly furnifhed at the end with a nail. Edges marked with fharp *Lamellæ*.
NOSTRILS, fmall, oval.
TONGUE, broad, edges near the bafe fringed.
FEET, middle toe the longeft.

Ex. *Br. Zool.* II. *tab.* XCVII.
Found in all climates.

Anas

SPANISH DUCK.

PATAGONIAN PINGUIN.

Anas of Linnæus, who has XLV fpecies. *Briffon* divides this genus into *Anfer* and *Anas*; has XVI of the firft, and XLII of the fecond.

Bill, ftrong, ftrait, bending a little towards the point.
Nostrils.
Tongue, covered with ftrong fharp fpines, pointing backwards.
Wings, very fmall, pendulous, ufelefs for flight; covered with mere flat fhafts.
Body, covered with thick fhort feathers, with broad fhafts, placed as compactly as fcales.
Legs, fhort and thick, placed quite behind.
Toes, four ftanding forward; the interior loofe, the reft webbed.
Tail, very ftiff, confifting only of broad fhafts.

Patagonian Pinguin, *Ph. Tr.* vol. LVIII. 91. tab. V.

Inhabits an ifle near the *Cape* of *Good Hope*, on the coafts of *New Guinea* *, the ifle of *Defolation*, fouth of the *Cape*, the fouthern parts of *South America*, and the feas among the ice as high as *fouth lat.* 64. 12. *long.* 38. 14. eaft †. Lives much at fea. The wings act as fins. On land burrows. Are analogous to Seals.

Diomedea demerfa and *Phaethon demerfus* of Linnæus. *Sphenifcus* and *Catarractis* of *Briffon* VI. 96. and 102. I call it Pinguin, the name firft given it by the *Dutch* voyagers, a *Pinguedine*.

Ex.

* Voy. de *Sonnerat*. 179.
† *Cook's* Voy. i. 38.

M

Bill,

XCIII.PELICAN. BILL, long and ftrait; the end either hooked or floping.

NOSTRILS, either totally wanting, or fmall, and placed in a furrow, that runs along the fides of the bill.

FACE, naked.

GULLET, naked, and capable of great diftenfion.

TOES, all four webbed.

Ex.   Pelican, EDW. XCII.

Corvorant, *Br. Zool. fol. tab. F.* 1. *Br. Zool.* II. *tab.* CII.

Congenerous birds, inhabit all parts of the globe.

Pelicanus of LINNÆUS, who has VIII fpecies.   *Briffon* divides
this genus into *Sula, Phalarocorax,* and *Onocrotalus,* and forms
out of them XII fpecies, VI. 494. 511. and 519.   All feed on
fifh.   The *Corvorants* fit, and often breed in trees.

XCIV. TROPIC. BILL, compreffed, flightly floping down.   Point fharp.   Under
mandible angular.

NOSTRILS, pervious.

TONGUE.

TOES, all four webbed.

TAIL, cuneiform: Two middle feathers extending for a vaft length
beyond the others.

Ex.   Tropic Bird, EDW. CXLIX.

Inhabits within the *Tropics.*   Flies very high.

Phaethon of LINNÆUS, from the great heighth it afpires to.   *Lep-
turus* of *Briffon,* from the flendernefs of its tail, VI. 479.

BILL,

PELECAN.

FRIGAT PELECAN.

BILL, long, ſtrait, ſharp pointed.

NOSTRILS.

TONGUE.

NECK, of a great length.

FACE and GULLET, covered with feathers.

TOES, all four webbed.

XCV. DARTER.

Black-bellied *Anhinga, Ind. Zool. tab.* XII.

Inhabits *Guinea, Ceylon,* and *South America.* Darts out its head either at its food, or at paſſengers that go by; whence the name.

Plotus of LINNÆUS, *Anhinga* of *Briſſon,* VI. 476.

Ex.

The intention of giving a plate to every genus of this work, was entirely ſuperſeded by the conſideration of a NEW ORNITHO-LOGY, undertaken by Mr. *John Latham,* of *Dartford.* In that comprehenſive attempt, every ſpecies of bird will be fully de-ſcribed; and one or more figures will be given, explanatory of each genus: Mr. *Latham*'s knowlege of the ſubjeᴄt, and the labor he has beſtowed on it, will doubtleſsly render it worthy of the attention of the public.

EXPLANATION

OF THE

P L A T E S.

Genus I. · F A L C O N.   I.

C R E S T E D   H O B B Y.

F. with black head, cheeks, and hind part of neck. Head
flightly crefted. Back, and coverts of wing, black. Primaries
and tail, of the fame color, marked with numerous bars of
white. Throat, white. Breaft, thighs, and vent, ferruginous.
Legs, yellow. Size of the *Englifh* Hobby.
From *Surinam*. Preferved in the *Britifh* MUSEUM.

Genus   V.   P A R R O T.   II.

W H I T E - C O L L A R E D   P A R R O T.

P. with a red bill; blue head, cheeks, and chin; green neck, back,
and wings. Neck half furrounded with a white collar, paffing
over the upper part towards the throat. Upper part of the
breaft

breaſt of a fine red; the lower, yellow: belly, blue: thighs, yellow and blue: tail, cuneated; yellow beneath.

Inhabits the iſles of the *Eaſt Indies?*

III.

Genus XXII. JACAMAR.

CUPREOUS JACAMAR, *fig.* 1.

ALCEDO GALBULA, *Lin. Syſt.* 182.

J. with a black bill: whitiſh throat: head, cheeks, wings, and tail, of a bluiſh green: breaſt, belly, and back, of a variable copper color, very rich and gloſſy.

Varies in ſome reſpects from that figured by Mr. EDWARDS, tab. XXXIII. in having leſs green on the back; ſo probably is of another ſex.

Size of a Lark.

Inhabits *Surinam*, and other parts of *South America*.

IV.

Genus XXVIII. CREEPER.

YELLOW-CHEEKED CREEPER.

Cr. with green head, back, wings, and tail: cheeks and throat, deep yellow: breaſt and ſides of a yellowiſh green, marked with bluiſh ſpots: belly, yellow.

Size, inferior by half to the *Engliſh* Creeper.

Inhabits *Surinam*.

Genus XXIII. KINGFISHER. V.

RED-HEADED KINGFISHER.

K. with a red bill; near the bafe of the upper mandible, a white
fpot: head, and upper part of the neck, of an orange red:
from each eye, towards the back, extends a purple line, termi-
nating in a white fpot; and on the inner fide of that, one of
black: chin, white: upper part of the back, a rich blue: the
lower, light purple: coverts of wings, black, edged with blue:
primaries, black: breaft and belly, yellowifh white: back,
orange: legs, red.

Size, leffer than the common Kingfifher.

Inhabits *India*.

Genus XXV. TODY. VI.

GREEN TODY, *fig.* 1.

TODUS VIRIDIS, *Lin. Syft.* 178.

T. with head, back, wings, and tail, of a fine green: throat, a
rich crimfon: breaft and belly, of a pale yellow: vent, deeper.

Size of a Wren. The Green Sparrow of Mr. *Edwards,* tab.
CXX.

Inhabits *Jamaica,* and the hot parts of *America.*

BROWN

B R O W N   T O D Y, *fig.* 2.

T. with the whole upper part of a ferriginous brown: the coverts
of the wings, croffed with a dufky bar: lower part of the body,
olive, fpotted with white: tail, ferriginous.

Size, larger than the former.

Inhabits the hot parts of *America.*

VII.

Genus   XXVI.   B e e ‑ E a t e r.

I n d i a n   B e e ‑ E a t e r.

M e r o p s   V i r i d i s, *Lin. Syft.* 182.

B. with head, and lower part of neck, of a fine light blue, bounded
below by a line of black: a black line paffes from bill through
the eyes to the hind part of the head: hind part of head and
neck, of an orange red: upper part of back, coverts of wings,
fecondaries, and tail, green: the middle part of the fecondaries,
of a reddifh orange: lower part of back, of a light blue:
breaft and belly, of a yellowifh green: tail, long; two middle
feathers two inches longer than the others, and appear like
mere fhafts.

Nearly the fize of a Redwing Thrufh.

Inhabits *India.*

VIII.

Genus   XXIX.   H o n e y ‑ S u c k e r.

Y e l l o w ‑ f r o n t e d   H o n e y ‑ S u c k e r, *fig.* 1.

H. with a yellow forehead: green body and coverts of wings:
black primaries and tail.

PURPLE‑

PURPLE-CROWNED HONEY-SUCKER, *fig.* 2.

H. with a purple crown: green throat: rich deep blue collar round the whole lower part of the neck: back, green: wings, and forked tail, of a deep purple.

ORANGE-HEADED HONEY-SUCKER, *fig.* 3.

H. with an orange head: yellow throat and breaft: deep brown belly and back: purple wings: bright ferruginous tail.
All very minute. Inhabitants of the hot parts of *America.*

Genus XLIII. CHATTERER. IX.

COTINGA.

AMPELIS COTINGA, *Lin. Syft.* 298.

Ch. with head and upper part of the body, and coverts of wings, of a moft fplendid blue, deepeft on the crown: belly and vent, of the fame color: under fide of the neck and breaft, of a lovely purple: in fome the breaft is croffed with a band of the fame blue with the upper part: primaries and tail, dufky.
Size of a Stare.
Inhabits *Surinam,* and other hot parts of *South America.*

N   Genus

X.

Genus LIII. MANAKIN.

CRESTED MANAKIN, *fig.* 1.

PIPRA RUPICOLA, *Lin. Syst.* 338.

M. with a whitish bill: great round upright crest, of a fine orange color, crossed near the end of each feather with a darker line: neck, back, and whole under side, of the same vivid color. The feathers on the back end singularly, as if they had been cut off: and some of the feathers on the sides of the back, are loose and pointed. The primaries brown, marked with a white line: tail short, partly brown, partly orange.

Size of a Turtle Dove.

Inhabits *Surinam.*

GOLDEN-HEADED MANAKIN, *fig.* 2.

PIPRA ERYTHROCEPHALA, *Lin. Syst.* 339.

M. with a rich yellow head: scarlet shoulders and thighs: black body, wings, and tail.

Size of a Wren.

Inhabits *Surinam.*

XI.

Genus LXVI. CURLEW.

PYGMY CURLEW, *fig.* 1.

C. with the head, back, and coverts of the wings, mixed with brown, ferruginous, and white: primaries, dusky, edged with
white:

white: breaſt and belly, and rump, white: tail, duſky: the exterior feathers edged with white: bill and legs, black.

Size of a Lark.

Inhabits *Holland.* Sent to me by Doctor L. Theodore Gro-novius.

Genus LXVIII. Sandpiper.

XII.

Little Sandpiper, *fig.* 2. *Br. Zool.* II. Nº. 207.

S. with head, upper part of the neck, back, and coverts of the wings, brown, edged with black, and pale ruſty brown: breaſt and belly, white.

Leſſer than a Hedge Sparrow.

Inhabits *Europe* and *North America.*

Genus XCI. Duck.

XIII

Spanish Duck.

Anas Vindila, *Lin. Syſt.* 205.

D. with forehead, cheeks, chin, and hind part of the head, of a ſnowy whiteneſs: crown, black: neck, ſurrounded with a black collar: back and breaſt, bright ferruginous, croſſed with nar-row duſky lines: wings, pale brown, without a ſpeculum: belly, whitiſh brown, ſpotted with black: tail, cuneiform, black: legs, bluiſh.

Size

Size of a Wigeon.   Whiftles like one.

Inhabits *Spain* and *Barbary*.   Prefented to me by Sir *Hugh Williams*, baronet.

XIV.   Genus   XCII.   P i n g u i n.

PATAGONIAN PINGUIN, *Ph. Tr.* Vol. LVIII. 91.

P. with a flender bill, flightly bending: head, hind and fore part of the neck, dufky: each fide of the neck marked with a narrow ftripe of bright yellow, pointing from the head to the breaft, uniting beneath the dark color of the front of the neck, and fading away into the pure white of breaft and belly: whole back, of a deep cinereous color, marked with numerous fmall cærulean fpots: outfide of wings, black: infide, white: tail and legs, black.

Length, about three feet three inches.   Weight, forty pounds*.

Inhabits *Falkland Iflands*, *New Georgia*, the ifland of *Defolation*, (vifited by Captain Cook, in his laft voyage, fouth lat. 48 $\frac{1}{2}$. eaft longitude from *Greenwich* 70.) and *New Guinea*.

The meafurement of that which I defcribed in the Philofophical Tranfactions, is faulty, being taken from a ftuffed fkin too much diftended.   I correct it here, and add a figure of the bird, taken from life.

* *Forfter*'s Voy. ii. 528.   To the places where the *Leonine Seals* are found, p. 535 of my *Hift. Quad.* add *Staten-land*, and the ifland of *Defolation*.

Genus

Genus XCIII. PELECAN.   XV.

COMMON PELECAN.

PELECANUS ONOCRATALUS, *Lin. Syſt.* 215.

P. with the head ſlightly creſted: plumage, white, tinged with a
fine bloſſom color: the primaries, black: legs and feet, pur-
pliſh: webs, duſky.

Size, ſometimes double that of a Swan: bill, twenty inches long:
extent of wings, eleven feet eight, *Engliſh* *.

Inhabits the *Danube*, in its paſſage through *Hungary*: abounds in
Africa and *Aſia*, in ſeveral parts: numbers about the *Caſpian*
ſea; and it is not unknown on lake *Baikal*. Found alſo of a
vaſt ſize in *New Holland*.

Genus the ſame.   XVI.

FRIGAT PELECAN.

PELECANUS AQUILUS, *Lin. Syſt.* 216.

P. with a ſlender bill, hooked at the end: under the lower mandible,
a vaſt naked ſcarlet pouch, like a bladder, extending down the
whole front of the neck. The uſes of this pouch to this and
the former ſpecies, is to convey food to their young: the common Pelecan, which often breeds in arid deſerts, makes it a ve-
hicle for water for its young brood. Color of the plumage of the
male, a deep brown, or chocolate; the coverts of the wings
lighteſt, and tinged with ruſt color: the longeſt feather in the tail,

* EDWARDS.

10   nineteen

nineteen inches long: the middle, or ſhorteſt, only eight: breaſt of the FEMALE, white.

Length, three feet.

Inhabits *Aſcenſion Iſland*, the *Weſt Indies*, ſome of the *Indian Iſlands*, and *Eaſter Iſland*, in the ſouth ſeas: hovers in the air with the gentle motion of a kite: feeds on fiſh: darts on its prey as the fiſh ſwim near the ſurface, but does not alight on the water: will aſſault other birds, and make them caſt up their prey; and then catch it in the fall *.

* *Dampier. Campeachy* Voy. 25.

E R R A T U M.

P. xxvi. l. 31, *for* Pelican, *read* Pelecan.

I N D E X.

I N D E X.

Printed in the United States
By Bookmasters